Exploding the Myths of Climate Change!

A Denier's Viewpoint

SECOND EDITION

Clifford Holliday

Clifford Holliday

Exploding the Myths of Climate Change!

Second Edition

A Denier's Viewpoint

January 2021, re-issued March 2021, Reissued March 2022

Copyright © 2021

Cover design by Alice Holliday.

All rights reserved. This includes all rights, national and international, to reproduce any part of this book in any form without the author's permission.

ISBN-13
979-8594123526 (paper back)

Nonfiction>Reference>Climate Change>Global Warming

This book is dedicated to all those in this world who have been, or will be, harmed by the misdeeds of the true believers in anthropogenic climate change.

A special dedication goes to my wife Alice, who has endured my ranting about climate change and never begrudged me the time to write this.

I also want to remember my many friends and relatives who assisted in editing this work.

This Second Edition adds information and opinion about possible fixes for the Texas 'Big Freeze' which occurred February 14-18, 2021. It is dedicated to those hundreds lost in that unnecessary disaster and to the suffering of millions of Texans. May this tragedy serve as a reminder of how serious these issues are. They affect the well being of every American.

Clifford Holliday

Table of Contents

EXPLODING THE MYTHS OF CLIMATE CHANGE! .... - 1 -

TABLE OF FIGURES .... - 8 -

FOREWORD .... - 10 -

INTRODUCTION .... - 12 -

Something about Global Warming/Climate Change Bothers Me .... - 15 -

CHAPTER 1: WHAT IS GLOBAL WARMING/CLIMATE CHANGE? .... - 18 -

The Logic Chain of Global Warming/Climate Change .... - 27 -

CHAPTER 2: THE COMPONENTS OF DISASTER: ALLEGED IMPACTS OF CLIMATE CHANGE .... - 30 -

All this Fuss over Less than a Degree Change in Temperature! .... - 32 -

CHAPTER 3: EXAMINING THE ALLEGED IMPACTS .... - 34 -

    Flooding .... - 36 -
    Drought, Famine .... - 38 -
    Increased Hurricanes .... - 41 -
    Climate Refuges .... - 44 -

Increased Insect-borne Diseases ....- 56 -
Acidification of the Seas ....- 60 -
Release of Methane from the Permafrost Melting ....- 68 -
The Glaciers Are Melting ....- 72 -
What about the Poles? ....- 75 -
Increasing Wildfires ....- 80 -
Species Extinction ....- 85 -
Climate Risk ....- 88 -
Summary of Effects ....- 92 -
2020??? ....- 94 -

CHAPTER 4: CAN AIR REALLY HEAT THAT MUCH WATER? ....- 100 -

Summary – The Broken Logic Chain ....- 103 -

A Mind Experiment ....- 104 -

CHAPTER 5: THINGS SHOULD CORRELATE: THEY DON'T! ....- 106 -

Summary – Broken Logic Chain ....- 112 -

CHAPTER 6: "97% OF THE SCIENTISTS AGREE." ....- 113 -

"97% of Scientists Agree on Global Warming." ....- 113 -
500 Scientists Disagree ....- 114 -
The Petition Project ....- 115 -
Meteorologists' Survey ....- 117 -
Summary of "97% agree" – They don't! .- 118 -

CHAPTER 7: "WHAT HAVE WE GOT TO LOSE?" ....- 120 -

What have we got to lose? Plenty! ....- 120 -
A Tale of Two Countries ....- 131 -
So What Have We got to Lose – a Lot! ...- 135 -
Confusion of issues with Climate Change- 137 -

CHAPTER 8: THE TEXAS BIG FREEZE- 138 -

What Have We Got to Lose? – The Big Freeze showed us. The Texas electrical grid is a perfect example..............................- 138 -
Investment Choices ....- 140 -
Base Load vs. Peaking ....- 142 -
So What Really Happened to Texas' Power Grid? ....- 144 -
Our Sins ....- 150 -
What are the Wages of Our Sins? ....- 151 -
How do We Fix This Mess? ....- 151 -
Texas – Summer of 2022 ....- 153 -
The Grid in 2022 ....- 154 -
'What if' Experiment – Texas Grid ....- 154 -

CHAPTER 9: DOOMSDAY PREDICTIONS THAT DIDN'T GO 'BANG!' – THE 'CRY WOLF' SYNDROME ....- 157 -

Al Gore ....- 158 -
AOC ....- 159 -
The Coming Ice Age! ....- 161 -
New York's West Side Highway Underwater by 2019. ....- 164 -
No More Snow ....- 164 -
Nostradamus Predicted the End of the World. ....- 165 -
The Mad Monk Predicted the End of the World ....- 165 -

Summary of 'Crying Wolf' ....- 166 -

CHAPTER 10: WITH ALL THESE BAD EFFECTS, HOW ARE WE DOING? - 167 -
- **Poverty Level** .... - 168 -
- **Crop Production** .... - 171 -
- **Bad Weather** .... - 173 -
- **Life Expectancy** .... - 174 -
- **Summary of Review of Our Real Current Situation** .... - 176 -

CHAPTER 11: EXPLODING THE MYTH OF CLIMATE CHANGE: SUMMARY - 177 -

EPILOGUE .... - 185 -

RECENT BOOKS BY CLIF HOLLIDAY . - 192 -

Table of Figures

Figure 1, Greenhouse Effect - 19 -
Figure 2, Greenhouse Gas Absorption - 20 -
Figure 3, Greenhouse Gases Relative Amounts and Effects - 23 -
Figure 4, Greenhouse Gas Increase - 24 -
Figure 5, Global Warming Logic Chain - 27 -
Figure 6, Detailed Logic Chain - 28 -
Figure 7, US Floods – Through 2012 - 37 -
Figure 8, Annual Rate of Deaths from Famine - 40 -
Figure 9, Land Falling Hurricanes - 43 -
Figure 10, Table of Worldwide Natural Disasters - 52 -
Figure 11, Table of Worst Natural Disasters since 1900 - 54 -
Figure 12, pH Scale - 62 -
Figure 13, CO2 and pH Measurements - 67 -
Figure 14, Arctic Daily Temperature - 2019 Compared to 1958-2002 - 77 -
Figure 15, Broken Logic Chain - 99 -
Figure 16, Broken Logic Chain - 103 -
Figure 17, Logic Chain - 107 -
Figure 18, CO2 Over Time Graph - 107 -
Figure 19, Temperature and CO2 Chart - 108 -
Figure 20, Broken Logic Chain - 112 -
Figure 21, Costs of Green New Deal - Estimate - 126 -
Figure 22, Global Warming Issues of Green New Deal - 127 -
Figure 23, German Energiewende Progress - 132 -
Figure 24, French Electricity Production - 2017 - 134 -
Figure 25, Peak vs. Base Load - 143 -

Exploding the Myths of Climate Change

Figure 26, Wind Energy and Peaking *- 144 -*
Figure 27, Power Sources in Texas by % *- 146 -*
Figure 28, Power Sources - Texas - MW *- 147 -*
Figure 29, Texas Power Generation by Source - Critical Period *- 148 -*
Figure 30, US Poverty Levels *- 169 -*
Figure 31, World Poverty *- 170 -*
Figure 32, US Corn Yield *- 171 -*
Figure 33, Selected Countries' Agricultural Production *- 172 -*
Figure 34, Deaths from Bad Weather *- 174 -*
Figure 35, US Life Expectancy *- 175 -*
Figure 36, World Life Expectancy *- 175 -*
Figure 37, Real Logic Chain *- 182 -*

Foreword

"…What are regular people (like me) suppose to believe about this climate change and global warming stuff? So many people deny it; others stand by it. They are all shaking graphs and screaming."

The following few points comprise an elementary guide for you. They don't cover every situation, but the points do give a head start evaluating who to pay attention to and who to ignore.

- · Do they have graphs starting in the 1900's? The later the chart begins, the more questionable the data. **Question these people.**
- · Do they have graphs starting in the prehistoric era? **Probably worth listening to, but I question them too.**
- · Do they give you a hard deadline? Such as "… are only X years from the tipping point (or disastrous flooding, or complete extinction, or any list of other things)?" **Ignore these people.**
- · Do they tell you to "believe the science?" **Ignore these people because science is not belief-based.**
- · Do they tell you some super-majority agrees with the impending doom? **Ignore these people**. 97% of people believed

- the earth was the center of the universe at one time too.
- · Do their actions match their claims? Have they purchased waterfront property while telling you about the dangers of sea-level rise, for example (like some prominent politicians?) **Ignore these people.**
- · Do they benefit personally from their proposals? Who gains financially from their recommendations? That doesn't make the plan wrong, but it helps to understand motives in promoting any particular set of actions. **Probably ignore these people or at least be extremely suspicious**.

This list may leave you ignoring an awful lot of the discussion about global warming/climate change. That's good, because an awful lot of it should be ignored. As I said, this list doesn't cover every possibility, but it should allow you to put yourself on substantial ground without much work.[1]

If you want to understand why these points are essential and what the fallacies of the climate change true believers are, this book is for you. "Exploding…" spells out the top climate change myths and destroys them. It will give you the entire picture. Sorry, but a few parts of it do have a lot of graphs and are a little technical. However, most of it is plainly written to get the points across, without any screaming.

---

[1] The idea for these points came from Chip Kerr November 12, 2020. Published in Quora.com.

Introduction

This is a book of mythology. Mythology is a collection of myths of a particular culture or religion, as in 'Roman Mythology,' or 'pagan mythology.' Since ancient times mankind has built systems of myths to explain the complicated world he didn't understand. These mythologies were dauntingly complex and involved many human-like relationships and relationships of an eccentric character. The whole system was deeply ingrained in the minds of the populations. For example, the Greeks attributed a season of poor crops to a disturbance caused by the displeasure of Ceres (also called Demeter), the goddess of agriculture and grains. Everything that happened adversely to these people was the fault of some divinity of the mythology.

Similar mythology has been built around climate change. A compound set of tales has been constructed to support the idea of human-caused climate change. This mythology, supporting its ridiculous claims and dicta, has come from the distortion of science. This distortion was achieved mainly by the media and those with other objectives (see, for example, the chapter on the 'Green New Deal'). Almost anything wrong in our world is attributed to a disturbance of the human-caused (anthropogenic) climate change mythology – droughts, floods, storms, fires, extinctions, diseases, etc. – all the fault of climate change!

In this book, I will try to separate the many climate change myths from the truths.

Exploding the Myths of Climate Change

Being called a 'denier' is not pleasant. It sounds bad. It seems mean. That, however, is what I am. Anyone who dares to disagree with the idea of 'global warming' (now morphed to 'climate change') is a denier. Deniers will often be called many worse things by supposedly serious scientists, activists, and left-wing hangers-on, who are the 'true believers' of climate change/global warming. I guess I can live with this appellation; I've been called worse. However, this one bothers me because of global warming/climate change – **a fanatical political belief, almost a religion**. I feel it can do extreme harm to our country through political processes supported by our media, some politicians, and even our schools. Now (2021), we have a president who professes to believe climate change is the biggest single threat to our existence. The capacity for seriously injuring our country is almost unlimited.

In no way do I deny the globe is getting warmer, the amount of CO_2 is increasing, and the seas are rising. All of these things are happening – they are measurable facts. It is the causes, impacts, and suggested 'remedies' of these things I debate. Confusion of correlation with causation bothers me, and this is one of the true believers' significant sins. My only agenda is to prevent the true believers of the climate change religion from replacing our paradise with their version of hell. (Of course, we recognize the hell that has been created by the Covid-19 virus, but the possibilities of false 'corrections' of climate change far outweigh even that disaster on a long term basis.)

Before we begin about climate change, I would like you to know a little about me to better understand my position. I am an

Clifford Holliday

engineer (electrical, Professional Engineer in multiple states) by training, not a 'climate scientist.' I hold a graduate degree in Business Administration and have authored and published many (over 50) technical/marketing, major research papers (and several books), thus demonstrating my research capabilities. You can read more details about me at the back of the book. I just want it known I don't claim to be a 'climate scientist,' whatever that is. But I have sufficient training in the physical sciences to fully understand the physics and chemistry behind the true believers' arguments. After all, much of this 'science' is relatively elementary. My business training also equips me to recognize a 'snake oil salesman,' and that ability is more used in studying this field than knowledge of the physical sciences. Also, it should be clear; I have no connection of any type to any energy company.

Politically I am a conservative, a Republican. That doesn't mean I adopt all of the conservative viewpoints. However, I do choose many of them, particularly the fiscal policies. While the idea of global warming/climate change has always bothered me, it has never been for political reasons. Denying observable facts such as the increase in temperature and CO2 would be foolish. Those are truths. However, I know from my other work, things that seem to go together often don't. Temperature rise and increasing CO2 may be a set that seems to go together, but it doesn't. That the rising CO2 solely causes temperature increase and rising CO2 is harmful are myths.

The whole idea of anthropogenic climate change has always seemed to me just a bit arrogant. Believing humans can 'accidentally' influence something so vast and complex as the climate is dauntingly presumptuous. It sounds like something I may have read in my younger days when I was fascinated with science fiction – maybe a 'planet builder' or similar.

Something about Global Warming/Climate Change Bothers Me

After considering many of the issues, I am even more bothered now with our new president about this litany we hear, maybe particularly from the younger true believers, who have created a religion out of climate change. **Everything that is wrong in the world or their lives is the result of climate change - just like ancient mythologies.** The list of things blamed on climate change is almost endless (and we shall see a lot of it is, at best, baseless; at worst, vicious.) As a part of blaming everything on climate change, the true believers conflate things having nothing to do with climate change and the rising levels of CO_2.

In recent political activities, we see how this can become a severe problem exemplified by introducing the 'Green New Deal,' much of which has nothing to do with climate change. However, the Green New Deal, which is in reality a government takeover of a significant part of our economy, is being sold as an attack on the 'causes' of global warming. As an aside from this book's purposes, why would anyone want the government to take over anything? This blatant nonsense was adopted by several of the Democratic candidates for President in 2020! What would happen if we actually elected one of them? Unfortunately we are finding out. Mr. Biden, the new Democratic President, has accepted the Green New Deal, and so, presumably, is a part of the cabal.

While we are blaming everything on climate change, actual causes can go undiscovered and uncorrected. The terrible fires in Australia in 2109 are a perfect example. Even though the temperatures were nowhere near an Australian record (set years ago), the wildfires were immediately blamed on high temperatures resulting from climate change. Only after the fires burned and spread for weeks, it came to light they were the result of many arson acts going on while the true believers were wringing their hands about climate change.

We have even been subjected to a singer at the 'Golden Globes' bemoaning in tearful tones a 'continent was on fire' from global warming! **It was arson lady!**

The California wildfires of 2020 are much the same story, except, in this case, a large part of the problem is forest mismanagement and poorly maintained utility lines and protection. Yes, of course, the drought conditions added to the problem, but California has had droughts for centuries. (Droughts are treated individually later in the book.) California has also been having extensive fires for centuries. Before the gold miners came in 1849-50, the Indians burned off large chunks of forest every year. This is nothing new. Forest management and utility conditions continue to worsen (largely due to California's drive toward 'clean' energy), making them more significant. These causes

Exploding the Myths of Climate Change

are driven by other true believers who won't allow a tree to be touched in the forest or put onerous regulatory burdens on the utilities causing them to skip routine maintenance.

We will describe climate change, list those things that are the alleged results of climate change, and compare the allegations to real life. Then we will look at other things about the concept that concern me. These will include a discussion of causation versus correlation. We will also examine the thermodynamic properties of air and water, and how those properties will not support a primary tenet of global warming. Finally, we will look at some similar myths that have been perpetrated throughout history. Perhaps on that trip, you can come to see the supposed '97 % of scientists who agree with climate change' may be wrong, or, more accurately, does not exist.

The 'true believers' of climate change may come to concern you, too.

Chapter 1: What is Global Warming/Climate Change?

Let's start by trying to understand what is meant by 'Anthropogenic Global Warming.' This term, tossed around a lot by the global warming faithful, refers to global warming/climate change caused by man's activities. The prime event cited by the true believers causing climate change is the release of carbon dioxide, CO_2, by man's actions. This release is mainly from the generation of electricity from fossil fuel (oil, coal, natural gas) and internal combustion engines for transportation and industry. The increase in such usage since about 1950 is often of particular interest to true believers. While these are the true believer's primary targets, some go much further, even wanting to stop the livestock industry because of the gas, cattle, and other livestock expels. (This idea fits nicely with a vegetarian bent.)

The global warming/climate change chain of logic is essential to understand. First, you must know CO_2 is a 'greenhouse gas.' This reference means it acts in the atmosphere to re-reflect a portion of the light (heat) from the Sun back to Earth, which has been first reflected from Earth out towards space. This added light tends to heat the atmosphere and, thus, the 'Greenhouse effect.' If it weren't for this effect, our planet would be much colder and probably would not support life, at least not life as we know it. So the greenhouse effect is a good thing, an excellent attribute.

The following figure shows this effect.

Figure 1, Greenhouse Effect [2]

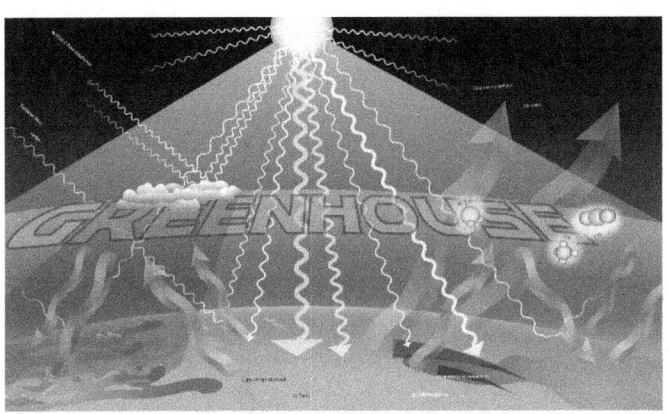

By A loose necktie - Own work, CC BY-SA 4.0, https://commons.wikimedia.org/w/index.php?curid=7833 6181

The small circles in this figure are molecules of CO2 shown re-reflecting heat (some go out in space, and some come back to Earth. The sketch doesn't show, but the next figure on absorption does show terribly well: how vital water vapor (clouds, etc.) is in this process.

The following figure shows the absorption of the light rays (shown for different wavelengths from the ultraviolet out to the far-

---

[2] A Loose Necktie, October 2019, "IPCC AR4 SYR Appendix Glossary" (PDF). Retrieved 14 December 2008

infrared) for the various greenhouse gases. Water vapor is the most important of all the gases; CO2 is second, as can be seen by the relative amount of absorption and scattering in the lowest block 'Major Components' in this figure.

Figure 2, Greenhouse Gas Absorption[3]

By Robert A. Rohde for the Global Warming Art project

---

[3]Rohde, Robert A., Creative Commons, Attribution-Share Alike 3.0 Unported, by for the Global Warming Art project

"This figure shows the absorption bands in the Earth's atmosphere (middle panel) and the effect this has on both solar radiation and up going thermal radiation (top panel). The individual absorption spectrum for major greenhouse gases plus Rayleigh scattering is shown in the lower panel.

"Both the Earth and the Sun emit electromagnetic radiation (e.g., light) that strictly follows a blackbody spectrum, and which can be predicted based solely on their respective temperatures. For the Sun, these emissions peak in the visible region and correspond to a temperature of ~5500° K. Emissions from the Earth vary following variations in temperature across different locations and altitudes, but always peak in the infrared.

"The position and number of absorption bands are determined by the chemical properties of the gases present. *In the current atmosphere, water vapor is the most significant of these greenhouse gases*, followed by carbon dioxide and various other minor greenhouse gases. Besides, Rayleigh scattering, the physical process that makes the sky blue, also disperses some incoming sunlight. Collectively

> these processes capture and redistribute 25-30% of the energy in direct sunlight passing through the atmosphere. By contrast, the greenhouse gases trap 70-85% of the energy in up-going thermal radiation emitted from the Earth's surface."[4]

From the above quotation it should be obvious there are several greenhouse gases plus water vapor, always entering and exiting the atmosphere. Water vapor (through evaporation, rain, and snow) is by far the largest by volume. Other than water vapor, CO2 is the most prominent greenhouse gas.

On a slightly more technical level, greenhouse gases have vibrational absorption bands (wavelengths), and non-greenhouse gases do not. Therefore the most abundant constituents of the atmosphere (nitrogen and oxygen) are not greenhouse gases as they don't have vibrational bands (at least they don't have vibrational bands that cause radiation).

From a study quoted in Wikipedia,

> "Schmidt *et al.* (2010) analyzed how individual components of the atmosphere contribute to the total greenhouse effect. They estimated water vapor accounts for about 50% of Earth's greenhouse effect, with clouds contributing 25%, carbon dioxide 20%, and the minor greenhouse gases and aerosols accounting for the remaining 5%."[5]

---

[4] ibid
[5] Schmidt, G.A.; R. Ruedy; R.L. Miller; A.A. Lacis

The following chart shows the relative concentrations of the various greenhouse gases in the atmosphere and the estimated total greenhouse impact from each.

Figure 3, Greenhouse Gases Relative Amounts and Effects[6]

(2010), "The attribution of the present-day total greenhouse effect" (PDF), J. Geophys.
[6] Kiehl, J.T.; Kevin E. Trenberth (1997). "Earth's annual global mean energy budget" (PDF). Bulletin of the American Meteorological Society. 78 (2): 197–208. Bibcode:1997BAMS...78..197K. doi:10.1175/1520-0477(1997)078<0197:EAGMEB>2.0.CO;2. Archived from the original (PDF) on 30 March 2006. Retrieved 1 May 2006.

Compound	Formula	Concentration in atmosphere (ppm)	Contribution (%)
Water vapor and clouds	H_2O	10,000–50,000	36–72%
Carbon dioxide	CO_2	~400	9–26%
Methane	CH_4	~1.8	4–9%
Ozone	O_3	2–8[B]	3–7%

notes:

[A] Water vapor strongly varies locally
[B] The concentration in the stratosphere. About 90% of the ozone in Earth's atmosphere is contained in the stratosphere.

By Kiehl, J.T.; Kevin E. Trenberth (1997).

Let's show, now, how (not necessarily 'why') the increase in CO2, which is a crucial part of the logic chain, is happening. The following graphs illustrate that point;

Figure 4, Greenhouse Gas Increase[7]

---

[7] *"Full Mauna Loa CO_2 record"*. Earth System Research Laboratory. 2005. Retrieved 6 May 2017.

Exploding the Myths of Climate Change

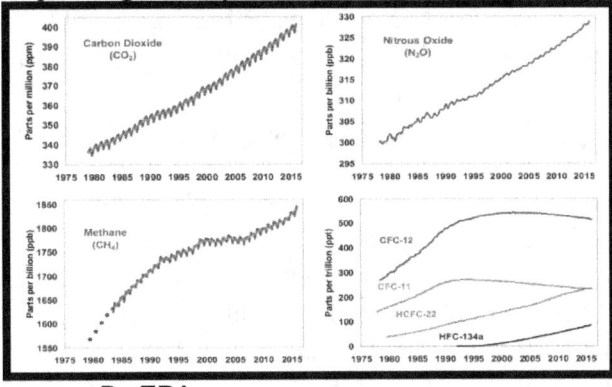

By EPA

This set of graphs illustrates all the major greenhouse gases (except water vapor) and how they increase through 2015. The figure on CO2 (upper left) shows how CO2 has been growing for the last 45 years steadily upward. The scale is essential in these graphs. For CO2 (carbon dioxide), the scale is in parts per million. So it shows CO2 going up to about 400+ parts per million in 2016. That is approximately .04% of the total atmosphere - much less than1%! N2O and CH4 scales are in parts per billion (that's 'B'), and the other chart of refrigerants (the 'F' compounds) are in parts per trillion. So even with the significant increases, we are not talking about much of the atmosphere.

Not shown on these charts is the most prevalent greenhouse gas – water vapor. Its concentration is estimated at between 10 and 50,000[8] parts per million. It is highly variable by location, as we all know by just considering it. The deserts are characterized by dry air (water vapor down near ten ppm), while the rain

---

[8] "Climate Change Indicators: Atmospheric Concentrations of Greenhouse Gases". Climate Change Indicators. United States Environmental Protection Agency. 27 June 2016. Retrieved 20 January 2017.

forests may be even higher than the 50,000 parts per million. A portion of the water vapor is composed of clouds, and some just condensation in the air. Except for the driest places (deserts), it will average about 4% of the atmosphere or about 100 times the concentration of CO2.

The reader may ask, "If water vapor is the most prominent greenhouse gas, why do we hear so much about CO2 and nothing about water vapor?" This would be an excellent question and one for which the true believers have a ready answer. The answer goes something like this: Since water comes and goes in the atmosphere (rain, snow, etc.), it is not as crucial as the longer-lived greenhouse gases (some CO2 can stay for many decades, even though it is dissolved in the sea, and used by plants.) Also, the true believers dismiss water vapor and usually never even include it in any discussion because they say it is a 'positive feedback mechanism' for other greenhouse gases. They mean if CO2 causes some warming, the added water vapor from warming will raise the level of water vapor (warmer air can hold more water) and thus increase even more the impact of the CO2 – positive feedback.

These points are correct, at least somewhat valid, but one must ask, "So what?" If water vapor is as vital as many say, and which even the 'rebuttal' seems to confirm, why is it never mentioned? Also, one would wonder if the feedback amount is a doubling of the impact of the CO2. Does this mean the amount of temperature rise due to increased CO2 in the last 70 years is only about .22°C?[9] That

---

[9] Frank, James, "Explaining How Water Vapor Greenhouse Effect Works,", 2015, Skeptical Science.

Exploding the Myths of Climate Change
would be a small amount to cause much concern.

I suggest the answer is: CO2 is needed as the culprit to give a rationale for all the radical changes the true believers advocate. After all, if the culprit is water vapor, what do you do about it? Where is the huge campaign to do away with fossil fuels, to 'keep it in the ground,' to stop eating meat, and so on with all the other drastic measures would completely change our way of life?

The Logic Chain of Global Warming/Climate Change

To begin our talk of climate change, we need to establish just what we are discussing. The global warming/climate change idea has been built on an exceedingly narrow set of logical (or maybe illogical) steps. It is that chain we are discussing. We are not debating plastic in the sea; we are not talking about air pollution (except as that is interpreted to mean CO2); we are not examining the quality of drinking water; in general, we are not reviewing the many causes sometimes are lumped into an 'environmental bag' and labeled 'climate change.'

The true believers' faith in global warming/climate change results in a particular chain of logic. It goes like this:

Figure 5, Global Warming Logic Chain

Fossil Fuels ➡ CO2 ➡ Air Heated ➡ Sea Heated ➡ Disaster!

- 27 -

To further explain, the following figure is offered.

Figure 6, Detailed Logic Chain

- There is an increased amount of CO_2 in the atmosphere due to the increased use of internal combustion engines and using fossil fuels (coal, oil, and natural gas) for power and heat generation. Besides, other activities of humanity (deforestation, raising cattle, etc.) are contributing lesser amounts to the CO_2 increase.

- The CO_2 is a greenhouse gas, and its increase means more light (heat) is reflected back to Earth rather than being radiated out to space.

- The increased back-reflection raises air temperatures.

- The rising air temperatures cause the seas to heat and raise their temperature.

Exploding the Myths of Climate Change

- The rising temperatures of the air and sea cause all of the catastrophic impacts listed in the next Chapter (except the Black Sox scandal.[10])

DISASTER!

Summarizing: Global warming/climate change is allegedly occurring because humankind is using more fossil fuels > adding more CO2 to the atmosphere > causing more heat to be reflected to Earth > raising the temperature of the atmosphere > rising the temperature of the seas > creating all kinds of disaster.

---

[10] For my non-sports minded readers, the Black Sox scandal occurred in 1919 when the Chicago White Sox (MLB) were accused of intentionally losing the World Series.

Chapter 2: The Components of Disaster: Alleged Impacts of Climate Change

Global warming/climate change has been blamed for a host of ills, some occurring, and some projected to occur. This is without the previously mentioned conflating of climate change with other environmental causes. Those on this list are all claimed to be the result of climate change. As an example of such claims, here is an excerpt from a book review appearing recently in my local newspaper (The Dallas Morning News):

> "... A more likely worst-case set of circumstances would include widespread crop failure and loss of water resources; an upsurge in extreme weather events such as floods, hurricanes, and heat waves; and an increase in tropical diseases. The results would be mass migrations and the possibility of conflicts that threaten life as we know it."[11]

Let's look at a list of these impacts as declared by the true believers, some of which are verifiable and, being magnanimous, we agree with verifiable facts.

The agreed on facts first:

---

[11] Bortz, Fred, "Debate on Climate Simmers," Dallas Morning News, 03-22-2006.

Exploding the Myths of Climate Change

Warming of the Atmosphere – Yes! This warming has occurred in that our current (2018) average temperature is <u>about .65° C or 1.17° F</u> [12] <u>above what it was in 1950.</u> The questions are, will it continue, and what caused it? In the period 1998-2007, the average temperature decreased[13], although some authorities dispute this decrease.

Warming of the Oceans – Again Yes! The oceans have warmed by between <u>.36° C and .52° C (average about .45° C or .80° F) in the last 40 years.</u>[14]

Rising Oceans – Yes! According to the European Environmental Agency, the oceans have risen by <u>20 cm since 1900</u> with significant variability from year to year and decade to decade.[15]

And now some less obvious and more contentious claimed impacts:

---

[12] Lewis, Marlo Jr. "Climate Change, Fossil fuels, and Human Well Being," July, 2018, Competitive Enterprise Institute.

[13] Clark, Josh, How Stuff Works, Are the Climate Skeptics Right?

[14] The International Panel on Climate Change (IPCC) Fifth Assessment Report

[15] Global and European Sea Level, European Environmental Agency, 2017.

Clifford Holliday

Flooding (other than Rising Oceans)

Droughts, Famine

Increased Hurricanes

Climate refugees

More Insect-born Diseases

Acidification of the Seas

Release of Methane from the Permafrost Melting

Glacial Melting

Fires

Species Extinction

> *All this Fuss over Less than a Degree Change in Temperature!*

All of this is over a <u>temperature differential of about .65° C in 70 years</u>.[16]

---

[16] Lewis, Marlo Jr., "Climate Change, Fossil fuels, and Human Well Being," July, 2018, Competitive Enterprise

Exploding the Myths of Climate Change

Considering differences in measuring techniques, locations, and types of measuring stations, etc. (measuring the temperature of the Earth is not as simple as measuring the temperature of an oven, which is not always so simple) is a tiny amount for all the concern.

This impressive list seems almost to have included opposites – floods and droughts? It has everything on it except alleging blame for the Black Sox scandal. There is no doubt the Earth has been and is warming; there is no doubt the seas are warming, and, yes, the waters are rising – and have been for many decades, maybe centuries. However, again only by a tiny amount. So tiny, that pictures of the same harbor, bay and island from all over the world taken many decades apart show no sign of a rising sea. As always, the devil's in the details.

Institute.

Chapter 3: Examining the Alleged Impacts

Myth: Global Warming/Climate Change is causing floods, storms, droughts, famines, species extinction, rising seas, melting glaciers, to get worse.

Truth: These things in fact are lessening. The worst cases of most of them occurred decades or centuries before CO2 became an issue.

Here is the first thing about climate change that bothered me. As noted and shown in Chapter 1, CO2 buildup has been occurring at least since 1950. It nearly has doubled since that time. The average temperature has also increased slightly in that time. Given the continuing CO2 increase (the 'demon' of the true believers); one would expect to find ample evidence of worsening occurrences of these alleged impacts of climate change. Also, one would expect a steady pattern of increase **if, in fact, these are the results of climate change.**

As I opened my research on the alleged disasters resulting from climate change for this book, I looked for the correlation between increased CO2 and more and worse disasters. I was looking for **cause and effect**. I found, **"That ain't so, Mac!"**

In looking at each of these different types of events (floods, famine, etc.)

individually, I have found, in general, the worst of each kind of incident happened long ago, rather than recently, as would be expected with a continually worsening CO_2 situation. Besides, **there is no discernible pattern to suggest they are getting stronger or more frequent.** Are, then, these disasters the result of anthropogenic (caused by man) global warming resulting in climate change, or are they the continuation of natural events that have been going on for many, many years? **It seems the only evidence is for the latter.**

In looking at this listing of various kinds of disasters, please remember, and the author is well aware, **all emergencies are unfortunate.** That's why they are called tragedies. If you are in a flood, famine, storm, etc., it doesn't make any difference if it is the worst ever or more or less frequent – **it is just terrible and often deadly.** However, in this book, we are looking to see if the logic chain of increasing disasters follows an increase of CO_2 as the climate change true believers claim. In other words, we are looking to establish **the cause and effect** relationship argued for climate change.

Let's consider each of these alleged impacts. Remembering the logic chain, if the true believers are correct, this steady growth **should be accompanied by a steadily increasing occurrence of the events ascribed to climate change, e.g., flooding, drought, famine, and all the rest.** The following sections examine each of these to see if we are experiencing the logic chain's alleged results. I will follow the methodology of first giving a true believer's rant about the impact of climate change on a particular type of disaster, and then I will contrast that with the real data.

Flooding

Here, this is taken to include rivers and seas, but not flooding due to rising ocean levels. Ocean level increases will be covered under a separate topic. Here is an example of the claims from climate change true believers about climate change causing flooding.

> *"Climate change is breeding storms with heavier rainfall, flooding farms — such as this one, which grows cotton.*
> *A warmer world — even by a half-degree Celsius — has more evaporation, leading to more water in the atmosphere. Such changing conditions put our agriculture, health, water supply and more at risk.*
> *Picture a North Carolina cotton farm that's been around since 1960, with global average temperatures rising by roughly half a degree since it grew its first crop.*
> *The increased evaporation and additional moisture to the atmosphere has led to 30% more intense rain during heavy downpours in that part of the U.S."*[17]

As noted, CO2 levels have been consistently going up since at least 1950 in a

---

[17] Ilissa Ocko Senior Climate Scientist, Monika Barcikowska Postdoctoral Scientist: Climate, Environmental Defense Fund, 257 Park Avenue South, New York, NY 10010, 2019.

Exploding the Myths of Climate Change

reasonably normal way. So if we are getting these effects from the resulting rising temperatures, we would expect to see an increase in the number of and severity of floods as time goes on. To check this, we have investigated the worst floods in World history and US history. The Wall Street Journal[18] has created a list of the worst floods in US history. Of these, **only one** – Hurricane Katrina (2005 in mostly Louisiana) at number five was in this century. The worst flood was Galveston in 1900. Of the top twenty, only two – Katrina and Superstorm Sandy in 2012 at number 17 – were the **only two** in the 21st Century. It is recognized, of course, these all had to do with water from oceans, but none were from ocean levels rising. **Those dates don't fit well with the hypothesis that flooding is getting worse because of climate change!**

Figure 7, US Floods – Through 2012[19]

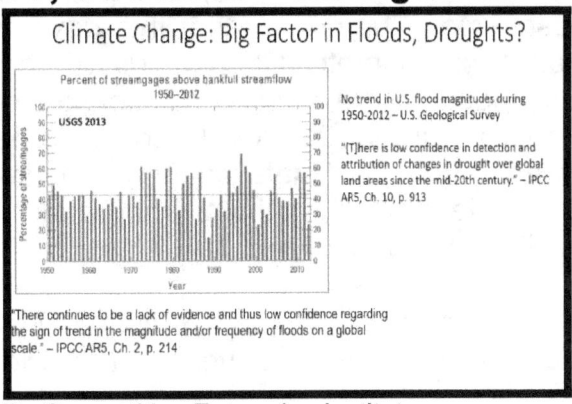

By Competitive Enterprise Institute

---

[18] Harrington, John, "Worst Floods in American History," September 2018, Wall Street Journal.
[19] Data from US Geological Survey; Comments from IPCC AR5.

Clifford Holliday

Looking past the United States to worldwide floods, Wikipedia[20] has created a remarkably informative list of flood disasters. The same pattern prevails in this list. Only the pattern is more pronounced. The first flood on the list in the 21st century is a China/North Korea flood in 2010, at number 41! There are only four on the list in the top 50 in the 21st century. **Again, this doesn't fit the true believer's claims of increasing floods from climate change.**

Drought, Famine

To me, the inclusion of drought and flooding seem to be citing opposites, but I guess they can be thought of separately. Famine seems to go with drought because the usual cause of famine (at least natural reason – excluding war and such) is drought. The fifth assessment report of the Intergovernmental Panel on Climate Change (IPCC) in 2013 concluded there was low confidence any significant trends in drought could be detected or attributed to climate change. However, the true believers still attribute drought and famine to climate change.[21]

> "Of all the ways climate change inflicts harm, drought is the one

---

[20] List of Deadliest Floods, Wikipedia. Deadliest floods worldwide with a minimum of 50 dead.
[21] IPCC, "2014: *Climate Change 2014: Synthesis Report.*" Contribution of Working Groups I, II and III to the Fifth Assessment Report of the Intergovernmental Panel on Climate Change [Core Writing Team, R.K. Pachauri and L.A. Meyer (eds.)]. IPCC, Geneva, Switzerland, 151 pp.

people worry about most, according to a Pew Research Center survey. And it's not surprising – droughts have been drier and lasting longer in recent years thanks in part to climate change. In 2012, the central and western US was hit particularly hard when 81 percent of the country was living in abnormally dry conditions, causing $30 billion in damages and putting the health and safety of many Americans at risk. ... While droughts can have different causes depending on the area of the world and other natural factors, the majority of scientists have started to link more intense droughts to climate change. That's because as more greenhouse gas emissions are released into the air, causing air temperatures to increase, more moisture evaporates from land and lakes, rivers, and other bodies of water. Warmer temperatures also increase evaporation in plant soils, which affects plant life and can reduce rainfall even more. And when rainfall does come to drought-stricken areas, the drier soils it hits are less able to absorb the water, increasing the likelihood of flooding – a lose-lose situation."[22]

---

[22] "The Facts About Climate Change and Drought," June, 2016, The Climate Reality Project,

With all the concern about drought and famine (which we will consider individually), one would expect drought was becoming a major worldwide problem. If the true believers' positions on climate change, causing all these problems due to the manmade increase in CO_2, are real, we would find data of recent serious issues increasing. We have looked at the data on drought **and find that not to be the case.**

The worst US drought occurred in the 1930's – the Dust Bowl. Since then, there have been several others in the US, but only one in the 21st century, in 2016, for a few months. In looking at worldwide droughts, the noted source lists the top twenty-five most devastating droughts. Of these, only three are in the 21st century. Many are in the 1900s (including the Dust Bowl) and many in centuries before. But the point is, there is **neither preponderance nor trend of recent droughts**, and the **worst droughts are long ago**, with the more recent ones being more modest (still terrible, but comparatively small.) **Again, nothing to support the claim climate change is causing more droughts and/or more severe droughts.**

Looking at famines, which are hard to separate from most droughts, except droughts do not always cause famines, we see a plainly definite trend. The following graph will illustrate the pattern.

Figure 8, Annual Rate of Deaths from Famine[23]

---

[23] Hasell, Joe and Roser, Max (2019) - "Famines". Published online at OurWorldInData.org. Retrieved from: 'https://ourworldindata.org/famines' [Online Resource]

Exploding the Myths of Climate Change

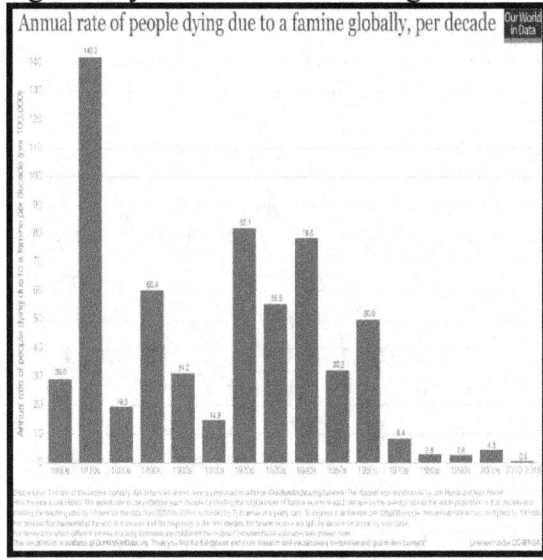

By OurWorldData.org

There is an evident lesson in the above chart – climate change is not causing deaths from famine. The table is a little hard to read, but each vertical bar represents worldwide deaths from starvation for the decade. There is an altogether certain trend of **dramatically decreasing deaths from this cause starting in the 1960s** and **actively continuing to today**. This finding **is just the opposite of what the climate change true believers are propounding.**

Increased Hurricanes

The true believers think increasing CO2 will cause an increase in the quantity and strength of hurricanes. The following quote is illustrative of the true believers' position on hurricanes.

> *"As global warming causes oceans to become warmer and more moisture is held in the atmosphere, the intensity of hurricanes and the amount of rain they produce will likely increase, according to NCAR scientist Kevin Trenberth and others. There is strong evidence global warming has been increasing the intensity of hurricanes for over the past few decades. ... The warming oceans are very likely causing the strength of hurricanes to increase. According to MIT scientist Kerry Emmanuel, hurricanes have become 70-80% more powerful over this time. Hurricanes take heat energy from the oceans and convert it into the energy of the storm. The warmer oceans offer more heat energy to hurricanes. This makes them become stronger storms."*[24]

In looking at the record, there is a mixed bag of actual results. The average number of significant storms (named hurricanes) hitting the US is 2.5 per year over a long period stretching back to the mid-1800. In the 21st century, there have been about as many years over that average as under it. There have been periods of relatively strong activity like 2016-17 and then periods of much weaker activity like 2018 and 2011-2015. There is certainly no

---

[24] Bergman, Jennifer, "Are Hurricanes Becoming Stronger and More Frequent?" February 15, 2011, Windows to the Universe.

Exploding the Myths of Climate Change
obvious trend to either more named hurricanes (the most occurred in 2020 – 13 and several years tied for the least – 0) or more intense storms.[25]

Figure 9, Land Falling Hurricanes

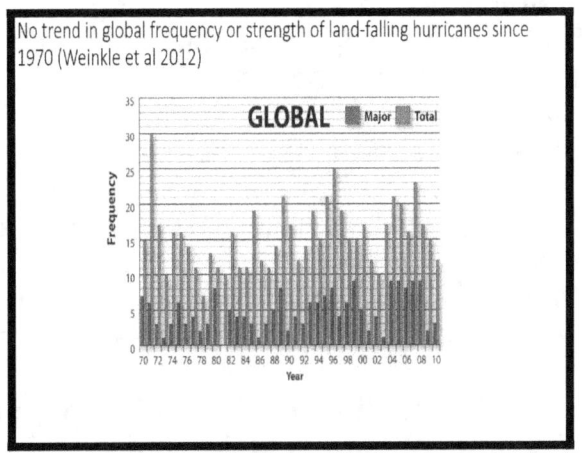

By Competitive Enterprise Institute, (Weinkle, et al, 2012)

While this is a mixed bag, **the trend that would support the true believer's hypothesis of climate change causing more storms and more intense storms is not there.**

---

[25] Landsea, Chris (NHC), 2019, Hurricane Research Division, Atlantic Oceanographic and Metrological Laboratory, National Oceanic and Atmospheric Administration. Note that the data was updated from here to reflect 2020 actual.

Clifford Holliday
Tornados

It is interesting to note we don't hear much about these extremely destructive storms that plague much of the southern and central United States annually. A single tornado seldom will do as much damage as a large hurricane, but tornados tend to come in bunches, and a given outbreak can be as destructive over a wide range of geography as nay hurricane. Perhaps the reason why we don't hear much about this is 2020 has been a relatively quit year for tornados.[26] As the reference shows the number of tornados impacting North America through December 1, 2020, is only a little more than half the maximum number, and close to the minimum. Why hasn't climate change/global warming caused an increasing quantity and severity of tornados?

Again, **the true believer's <u>hypothesis of climate change</u> causing more storms and more intense storms <u>is not there</u>**.

Climate Refuges

In a way, this is a little harder to tackle than something as concrete as 'flooding.' Either you had a flood, or you didn't. However, it is harder to define 'climate refugees,' unless you are a politician, then the definition doesn't matter – it's just wrong! Are the streams of people, 'climate refugees, and 'weather refugees, only refugees, or the latest politically correct moniker'climate migrants?' Maybe it would be better if we spent less time trying to

---

[26] NOAA, Storm Prediction Service, Dec. 1, 2020.

name these unfortunate people in ways to make political points (for example, see some of the mob of contenders for the 2020 Democratic Presidential Nomination) and instead seriously tried to help them. (By the way, by helping them, I do not mean violating US or any other country's immigration laws.) The true believers are labeling these people climate refugees, or now climate migrants, and proclaiming the worsening climate, due to anthropogenic global warming, is at fault. The following is an example of some of the typical claims from true believers:

> *"In other words, we're at a very, very weird moment in the trajectory of climate change awareness. With many people already suffering from climate consequences and many, many more poised to join them, we must convince those in resource-chugging countries to take action without inflaming their, at times misinformed, sense of self-preservation. The scale of action that must be taken is both overwhelming and overdue, and it requires seeing ourselves as a global community. But it's an incredibly complicated thing to do, and we must choose our words wisely, as pedantic as that can seem.*
>
> *Now to the numbers part of your question: The Institute for Economics and Peace, an Australian think tank, recently estimated in 2017 alone,*

Clifford Holliday

> *18 million people — 61.5 percent of global displacements — were forced to move due to natural disasters. (Those natural disasters are not universally caused by climate change, but global warming is predicted to cause more frequent and intense disasters.) And while projections vary, sources agree those numbers will get a whole lot higher. That same report noted nearly 1 billion people currently live in areas of "very high" or "high" climate exposure, which could result in millions of people displaced by climate change in the future. A 2018 World Bank report estimated by 2050, there would be 143 million climate change-driven migrants from the regions of Latin America, sub-Saharan Africa, and Southeast Asia alone."*[27]

Note even in this true believer quote, they are admitting some of what they are calling climate change impacts are perhaps just natural disasters – **but global warming will (in the unknown future) make them worse.**

Here's another quote from a true believer:

---

[27] Andrews, Eve, "How many climate refugees are there?" Jun 20, 2019, Umbra, Grist.

Exploding the Myths of Climate Change

"Rising sea levels, wildfires, drought or a slow-moving hurricane can all threaten your way of life as surely as a bomb. But unlike victims of war for whom the causes and effects of the threats are clear and codified, there are no protections for environmental migrants. That's despite World Bank estimates tens of millions of people could be climate refugees by 2050, research showing asylum seeking already increases in response to climate-related issues like rising temperatures, and the fact researchers have been talking about the plight of people who must escape not-entirely-natural disasters since at least 1988.... Here and around the world, fleeing climate change means running to bureaucracies as inhospitable to your survival as the places you left behind."[28]

This piece points out the world's legal systems don't recognize 'climate refugees' and laments that fact. It also alludes to how global warming/climate change is making the situation even worse.

Here's one more quote from a true believer to get the full picture of how this problem of refugees is viewed from a United Nations agency.

---

[28] Koerth, Maggie, "The World Isn't Ready For Climate Refugees," Sept. 12, 2019, FiveThirtyEight.

"Research indicates the Earth's climate is changing at a rate that has exceeded most scientific forecasts. Some families and communities have already started to suffer from disasters and the consequences of climate change, which has forced them to leave their homes in search of a new beginning.

UNHCR recognizes the consequences of climate change are extremely serious, including for refugees and other people of concern. The Global Compact on Refugees, adopted by an overwhelming majority in the UN General Assembly in December 2018, directly addresses this growing concern. It recognizes 'climate, environmental degradation and natural disasters increasingly interact with the drivers of refugee movements.'

The impacts of climate change are numerous. Limited natural resources, such as drinking water, are likely to become even scarcer in many parts of the world. Crops and livestock struggle to survive in climate change 'hotspots' where conditions become too hot and dry, or too cold and wet, threatening livelihoods and exacerbating food insecurity.

People are trying to adapt to the changing environment, but many

Exploding the Myths of Climate Change

> *are being forcibly displaced from their homes by the effects of climate change and disasters, or are relocating in order to survive. New displacement patterns, and competition over depleted natural resources can spark conflict between communities or compound pre-existing vulnerabilities."* [29]

To offer the last opinion on this subject, we quote from a Brookings Institutes report purports to point out how climate change is making the migrant problem worse.

> *" While Cyclone Idai was the worst storm in Mozambique's history, the world is looking towards a future where these "unprecedented" storms are commonplace. This global challenge has and will continue to create a multitude of critical issues that the international community must confront, including:*
>
> *Large-scale human migration due to resource scarcity, increased frequency of extreme weather events, and other factors, particularly in the developing countries in the earth's low latitudinal band*

---

[29] Gangale, Riccardo, "Climate Change and Disaster Displacement," 2019, UNHCR – United Nations High Commissioner for Refugees,.

> *Intensifying intra- and inter-state competition for food, water, and other resources, particularly in the Middle East and North Africa*
>
> *Increased frequency and severity of disease outbreaks*
>
> *Increased U.S. border stress due to the severe effects of climate change in parts of Central America*
>
> *All of these challenges are serious, but the scope and scale of human migration due to climate change will test the limits of national and global governance as well as international cooperation."* [30]

When taking all these quotes together, they are all saying climate change/global warming is making disasters worse and more frequent. As a corollary, then increased disasters will make the refugee problem worse, and, presumably, since CO_2 has been growing for at least 70 years, we already see that. The question is how to evaluate these claims. Is global warming/climate change making more migrants? How do we separate climate change-related migrants from natural disaster migrants? Is that separation necessary? – Probably not to the migrants.

In thinking about these questions, it seems the cause of the natural disaster is insignificant. If these disasters are getting worse due to climate change and creating

---

[30] Podesta, John, "The Migration-Climate Nexus Is Real, but More Scrutiny and Action Are Required," 2019, The Center for American Progress.

Exploding the Myths of Climate Change

more migrants, we should research the history of disasters, not worrying about how much climate change has impacted each one. If climate change is causing a pattern of worse disasters, it should show up as more, worse disasters recently with an upward trend.

    To check this out, I have searched for the deadliest natural disasters and found the following. I will print the entire table so the reader is not dependent on my interpretations.

Figure 10, Table of Worldwide Natural Disasters

Rank	Death toll (estimate)	Event	Location	Date
1.	1,000,000–4,000,000[2][nb 1]	1931 China floods	China	July 1931
2.	900,000–2,000,000[3]	1887 Yellow River flood	China	September 1887
3.	830,000[4]	1556 Shaanxi earthquake	China	January 23, 1556
4.	≥500,000[2]	1970 Bhola cyclone	East Pakistan (now Bangladesh)	November 13, 1970
5.	316,000	2010 Haiti earthquake	Haiti	January 12, 2010
6.	300,000	1839 India cyclone[5]	India	November 25, 1839
		1737 Calcutta cyclone[6]	India	October 7, 1737
8.	273,400[7]	1920 Haiyuan earthquake	China	December 16, 1920

Exploding the Myths of Climate Change

Rank	Death toll (estimate)	Event	Location	Date
9.	250,000–300,000[8]	526 Antioch earthquake	Byzantine Empire (now Turkey)	May 526
10.	242,769–655,000	1976 Tangshan earthquake	China	July 28, 1976

Source – Wikipedia Note: This table does not include volcanic eruptions, acts of war (blowing up dikes, etc.), nor flooding (see separate heading in this section).[31]

In reviewing the table, one is struck that the only disaster from the 21st century making the list is an earthquake (one of the few things not attributed to global warming) in Haiti in 2010. Half of the worst disasters occurred before the 20th century. That is not a particularly good backing for the case that global warming (with CO_2 increasing dramatically since 1950) is causing worse disasters, and thus more 'climate migrants.'

To further evaluate this proposition, I have found a table that lists natural disasters since 1900, thus only looking at more recent tragedies, and perhaps finding the pattern of increasing disasters in recent times.

Figure 11, Table of Worst Natural Disasters since 1900

---

[31] "List of Natural Disasters by Death Toll," Wikipedia, 2019.

Rank	Death toll (estimate)	Event*	Location	Date
1.	1,000,000–4,000,000	1931 China floods	China	July 1931
2.	≥500,000[2]	1970 Bhola cyclone	East Pakistan (now Bangladesh)	November 1970
3.	316,000[9]	2010 Haiti earthquake	Haiti	January 12, 2010
4.	273,400	1920 Haiyuan earthquake	China	December 16, 1920
5.	242,769–655,000	1976 Tangshan earthquake	China	July 28, 1976
6.	229,000	Typhoon Nina—contributed to Banqiao Dam failure	China	August 7, 1975
7.	227,898	2004 Indian Ocean earthquake and tsunami	Indian Ocean	December 26, 2004
8.	145,000	1935 Yangtze river flood	China	1935
9.	143,000	1923 Great Kantō	Japan	September

Exploding the Myths of Climate Change

Rank	Death toll (estimate)	Event*	Location	Date
		earthquake		1, 1923
10.	138,866	1991 Bangladesh cyclone	Bangladesh	April 29, 1991

Note: This table does not include volcanic eruptions, acts of war (blowing up dikes, etc.), nor flooding (see separate heading in this section). [32]

With this listing of the worst natural disasters since 1900, we still have only two in the 21st century and they are both earthquake related – nothing to do with global warming. Only half of the remaining eight occurred in the second half of the 20th century. After this analysis, it is certain there is no pattern of worsening disasters from global warming or anything else in the current time. If there are more refugees (and I think this would be extremely hard to prove, because refugees have had many causes over history – mostly war probably), they must be caused by some other driver. **It doesn't appear the facts support any connection with global warming/climate change-driven events causing refugee problems. 'Climate Refugees' may be a convenient political concept, but not one that is analytically supportable.**

---

[32] ibid

Increased Insect-borne Diseases

Regarding insect-born diseases, the contention is the increased temperature due to global warming will improve the breeding grounds for many mosquitoes and other insects, and thus will cause an increase in diseases born by these insects. The following is a typical statement of that position posted on a Scientific American Web site.

> "Is there a link between the recent spread of mosquito-borne diseases around the world and environmental pollution?
>
> If by pollution you mean greenhouse gas emissions, then definitely yes. According to Maria Diuk-Wasser at the Yale School of Public Health, the onset of human-induced global warming is likely to increase the infection rates of mosquito-borne diseases like malaria, dengue fever and West Nile virus by creating more mosquito-friendly habitats.
>
> "The direct effects of temperature increase are an increase in immature mosquito development, virus development and mosquito biting rates, which increase contact rates (biting) with humans," she reports.[33]

---

[33] Diuk-Wasser, Maria Ana PhD, "Mosquito-borne Diseases on the Uptick—Thanks to Global Warming," September 27, 2013, EarthTalk.

Exploding the Myths of Climate Change

In following our same pattern, we have attempted to look at least for a connection between increases in these diseases in relatively recent times. Again, global warming's main driver – increasing CO_2 - has been going on at least since 1950. So as with our previous alleged global warming impacts, we look for an increase in these diseases born by mosquitoes.

In checking malaria first, I went to the World Health[34] site and found reported there in 2017 there were 219,000,000 cases of malaria worldwide. That is certainly a significant number. This number of cases resulted in 435,000 deaths. Again, that is a substantial and concerning amount. However, it was also reported this large number represented an 18% **decrease** since 2010 in the number of cases. There was a 28% **decrease** in the number of deaths in that same period. **These decreases are a contraindication for climate change causing increased disease.**

Perhaps another disease will show the pattern of increasing occurrences. Dengue Fever is such a disease with an increasing occurrence rate in the tropics, mainly Asia, but also in Middle and South America and Africa. This disease has grown to as much as 390,000,000 infections, a six-time increase since 2010. It has a shallow death rate, but there is no specific treatment for it. A difficulty with getting exact information on Dengue Fever is a vast majority of the infections are not medically treated, being minor with cold or flu-like symptoms.

The following material from the World Health Organization explains the difficulty of accurately assessing the quantitative

---

[34] "Global Health Observatory– Malaria," 2019, WHO.

Clifford Holliday
information about Dengue and the variability of this information:

> "The number of dengue cases reported to WHO increased ~6 fold, from <0.5 million in 2010 to over 3.34 million in 2016. These numbers are from member States in only three WHO regions (SEARO, WPRO and PAHO), who regularly report their case numbers; there are other countries and regions that do not provide reports. This alarming increase in case numbers is partly explained by a change in national practices to record and report dengue to the Ministries of Health, and to the WHO. But it also represents government recognition of the burden, and therefore the pertinence to report dengue disease burden. Therefore, although the full global burden of the disease is uncertain, this growth is only bringing us closer to a more accurate estimate of the full extent of the problem…. In 2017, a significant reduction was reported in the number of dengue cases in the Americas - from 2 177 171 cases in 2016 to 584 263 cases in 2017. This represents a reduction of 73%. Panama, Peru and Aruba were the only countries that registered an increase in cases during 2017.
>
> Similarly, a 53% reduction in severe dengue cases was also

> *recorded during 2017. The post Zika outbreak period (after 2016) has seen a decline of cases of dengue and the exact factors leading to this fall are still unknown. WHO's Western Pacific Region has reported dengue outbreaks in several countries in the Pacific, as well as the circulation of DENV-1 and DENV-2 serotypes."*[35]

All-in-all, while this disease does show a significant recent increase, it is apparently an outlier and only recently. It has not been increasing over the period from 1950 in any steady fashion. Its sudden spike in cases seems to be more related to an as yet unknown cause – not global warming. **WHO in discussing this increase has never indicated it was in any way related to climate change.**

West Nile is another disease associated with types of mosquitoes, and usually is notably mild (cold, flu-like) but can develop into a much more serious ailment. It was first detected in Africa on the West Nile (of course) in 1937. It has since spread worldwide on every continent except Antarctica. During the early 2000s, it was an inordinately severe disease in the United States (where the cases are required to be reported to the Center for Disease Control.) In the period from 2002 to 2009, the disease averaged almost 4000 new cases a year in the US. However, since that

---

[35] "Dengue and Severe Dengue," 2019, World Health Organization. All information about dengue was taken from this source.

time, it has fallen off considerably so in 2013-2018 it only averaged a little over 2000 new cases a year – **a decrease of 50%.** (Note 2012 was an exception with over 5000 cases reported – an outlier number.)

Again, while West Nile remains a significant health threat, **it does not show a pattern of steadily increasing incidents due to climate change (or anything else), as the true believers suggest. After looking at these three diseases, often cited by true believers, there is no evidence of a link to climate change as correlated to increasing CO2. Instead, the opposite is indicated!**

The most recent and most disastrous modern disease has been the 2020 pandemic of Covid-19. As yet, no one has blamed that on climate change. Just give them time – they will.

Acidification of the Seas

By the 'seas,' this heading means the oceans and relates to a claim made by true believers the oceans are being made more acidic because of climate change. This 'acidification' is an intensely technical subject receiving a lot of attention in the scientific community. If one searches ('Google's'), you might get the impression of this being settled science – just showing how deep-seated the climate change true believers are. You have to look for dissenting opinions, even though there are plenty.

For the completely non-technical reader or one who hasn't been in Chemistry 101 for many decades, 'acidity' refers to the proportion of H+ ions (hydrogen ions) present in a liquid as compared to distilled water. More H+ increases acidity and confusingly enough,

Exploding the Myths of Climate Change

lowers the pH. The pH is the standard measure of acid/alkali, and a pH of 7.00 is considered neutral, while a higher pH – over 7.0 – is considered bass, and less than 7.0 is considered acid. The higher or lower the number, the less or more acid. The following chart, taken from an EPA publication, [36] shows how this pH scale works.

The reader will note seawater is shown as having a pH of 8.1, which is a commonly accepted average for the oceans – **alkali or basic, not acid.** The pH scale is logarithmic, as shown in the figure. To go up one point in pH, say from 8 to 9 (more basic, less acidic), reduces the amount of H+ ions from 1/10 to 1/100 a 1000% decrease.

---

[36] "Understanding the Science of Ocean and Coastal Acidification," 2017, US EPA.

Figure 12, pH Scale

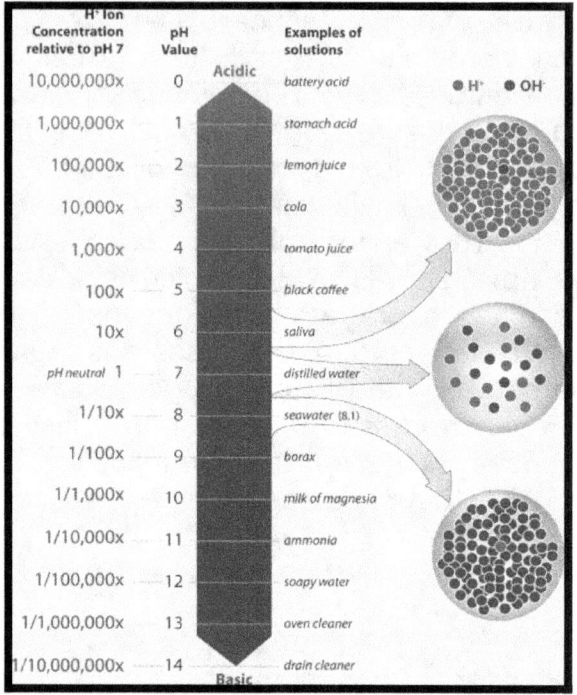

[37] By EPA

In any event, let's discuss what is meant by ocean acidification, its supposed impacts, and what is happening. Ocean acidification is happening due to the increase in CO2 in the atmosphere. The CO2 dissolves in the seawater, and through some chemical processes we will not detail, releases added hydrogen ions (H+), thus making the seawater less basic (more acidic.) This increase in acidity is said to cause 'bleaching' of the shells of some corals, and perhaps be detrimental to other marine life. Here is a quote illustrating the true believers' view:

---

[37] ibid

"Ocean acidification is sometimes called "climate change's equally evil twin," and for good reason: it's a significant and harmful consequence of excess carbon dioxide in the atmosphere that we don't see or feel because its effects are happening underwater. At least one-quarter of the carbon dioxide (CO2) released by burning coal, oil and gas doesn't stay in the air, but instead dissolves into the ocean. Since the beginning of the industrial era, the ocean has absorbed some 525 billion tons of CO2 from the atmosphere, presently around 22 million tons per day.

At first, scientists thought this might be a good thing because it leaves less carbon dioxide in the air to warm the planet. But in the past decade, they've realized this slowed warming has come at the cost of changing the ocean's chemistry. When carbon dioxide dissolves in seawater, the water becomes more acidic and the ocean's pH (a measure of how acidic or basic the ocean is) drops. Even though the ocean is immense, enough carbon dioxide can have a major impact. In the past 200 years alone, ocean water has become 30 percent more acidic—faster than any known change in ocean chemistry in the last 50 million years."[38]

There is no doubt this effect is happening. It is introductory chemistry. However, there are questions as to whether this is the whole story. Note how this quoted writer (and many other true believers when discussing this subject) refers to the 'ocean acidification' as if the seas have turned to acid. How horrible! We must stop that! We can't have acid seas!

Except, let's understand, this is <u>alarmist language chosen for its shock value.</u> It is highly deceptive, especially to an audience not trained in chemistry. They could have said, more accurately and less shockingly, "The seas are becoming slightly less basic." But that kind of language would not attract more true believers. **The seas are not acid. No one, not even the most devout true believer, is projecting they will become acid. The oceans are basic - alkali, not acid!** In the years since the industrial revolution, the seas have dropped their average pH value by an astonishing .1 (from 8.2 to 8.1 – don't ask me how they measured the average ocean pH in 1890 – it is still hard to do today and controversial.) The true believers are quick to point out, while this is a small amount, it represents a 25-26% increase in 'acidification' due to the previously explained logarithmic nature of the pH scales. Of course, this is true, but the **ocean is still basic, not acid, by a long shot**. Going from 8.1 pH to a neutral pH (7.0 - yet not acid) would require over a ten-point reduction or 1000% reduction!

---

[38] "Ocean Acidification," The Ocean Portal Team, 2019, Reviewed by Jennifer Bennett (NOAA), Smithsonian – Find Your Blue.

As a further example of how insensitive the sea is about pH with changes in CO2, the following experiment is cited.

> *At James Cook University's experimental aquarium facility, clownfish were sustained in non-manipulated seawater that obtained a pH of 8.15 ± 0.07 which is similar to our current ocean's pH. To test for effects of different pH levels, seawater was manipulated to three different pH levels, including the non-manipulated pH. The two opposing pH levels correspond with climate change models that predict future atmospheric CO2 levels. In the year 2100 the model predicts we could potentially acquire CO2 levels at 1,000 ppm, which correlates with the pH of 7.8 ± 0.05. Continuing even further into the next century, we could have CO2 levels at 1,700 ppm, which correlates with a pH of 7.6 ± 0.05.* [39]

As one further point on measuring the 'average ocean pH value' and its changes, the reader's attention is directed to the following set of graphs, again taken from the EPA showing measurements of CO2 and pH taken at various places.

---

[39] Introduction, in Zeebe 2012, p. 142

Clifford Holliday

There is an immense amount of information on these six graphs, but for this discussion, the reader's attention is directed to the top-right chart concerning the pH measured near Bermuda between 1983 and 2015. Note the swings in the measurements. Many of the variances are .1 acidity or more – the **amount the ocean is said to have increased in the last 100+ years**. Also, note some of **the latest 'high' measurements are extremely close to the earliest' high' ones**, even in this time frame. Interesting?

Here is also some more information of interest regarding the CO2 measurements. Note first, they vary significantly from sample to sample. Additionally, note the scale. If these charts were drawn on percentage changes, the movement of the pH would be infinitesimal. The CO2 is swinging by 25-33% while the pH is moving, maybe 2%. So while we have experienced considerable changes in CO2, there have been minimal changes in the ocean's pH, even if we concede this change can actually be measured.

Figure 13, CO2 and pH Measurements [40]

Exploding the Myths of Climate Change

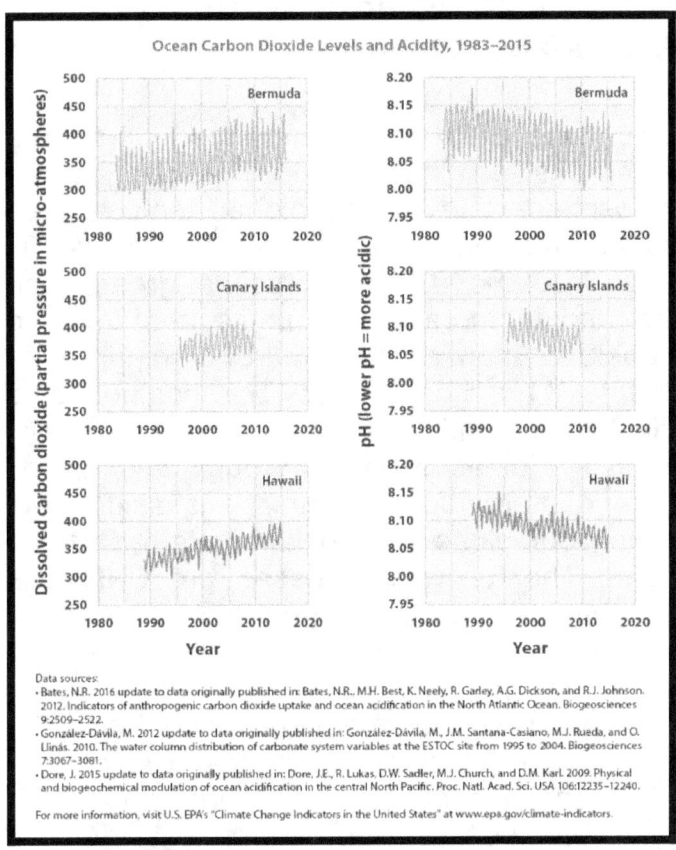

By EPA

In summary, regarding the change in pH (acidity) of the oceans, let's understand how much has occurred. **As with the warming of the air and the oceans, we find, in fact, inordinately little has happened**. The dire predictions are all based on what is 'projected' to occur. **The projections are incredibly speculative and of no demonstrable merit.**

---

[40] "Understanding the Science of Ocean and Coastal Acidification," US EPA, 2017

Clifford Holliday
Release of Methane from the Permafrost Melting

One of the effects claimed of climate change is an increase in methane in the atmosphere due to the higher temperatures induced melting of the permafrost in the Arctic area. Methane (CH_4) is itself a greenhouse gas and is more potent in its heat-trapping properties than the far more common (see Chapter 1) carbon dioxide. This compounding effect (the higher temperatures due to global warming releases more greenhouse gases causing more global warming) is said to be driving our globe toward higher temperatures faster. The following release from CNN tells how this is a climate change effect.

"Russian scientists studying Arctic waters found the most powerful ever methane jets shooting up from the seabed to the water's surface, they said Friday. Igor Semiletov, the chief scientist aboard a vessel carrying 65 scientists on a 40-day research voyage, told CNN via satellite phone he found amounts of methane in the air over the East Siberian Sea up to nine times the global average. Methane is a powerful greenhouse gas, with a significantly greater global warming potential than carbon dioxide, according to NASA. The methane emissions in the Arctic, fueled by the melting of

> *permafrost on the sea floor, are one driver of climate change, NASA said. The emissions are presenting a growing risk. Methane levels Semiletov's team found in the air above the seawater were "extremely high," he said. "Nobody has detected these concentrations." Levels are highest seen in decades of research Semiletov, a professor at Tomsk Polytechnic University in Siberia, said the ship full of scientists reached the East Siberian Sea around the beginning of October. The water is usually tough to get through due to it being "covered in ice," but Semiletov said this year was different. The water was "fully open."*"[41]

Note this describes the previously outlined compounding effect of higher temperatures leading to a more significant release of methane, leading to higher temperatures. **But this is not the whole story!** The following is a quote explaining how this is the current manifestation of a process started thousands of years ago. (This doesn't necessarily mean it is not a phenomenon that should concern us. It just means it is not caused by global warming.)

> "Hong et al. (2017) studied the seepage from large mounds of hydrates in the shallow arctic

---

[41] Prior, Ryan, CNN, October 12, 2019.

seas at Storfjordrenna, in the Barents Sea close to Svalbard. They showed though the temperature of the sea bed has fluctuated seasonally over the last century, between 1.8 and 4.8 °C, it has only affected release of methane to a depth of about 1.6 meters. Hydrates can be stable through the top 60 meters of the sediments and the current rapid releases came from deeper below the sea floor. They concluded the increase in flux started hundreds to thousands of years ago well before the onset of warming others speculated as its cause, and these seepages are not increasing due to momentary warming.[17] Summarizing his research, Hong stated:

"The results of our study indicate the immense seeping found in this area is a result of natural state of the system. Understanding how methane interacts with other important geological, chemical and biological processes in the Earth system is essential and should be the emphasis of our scientific community,"]

Further research by Klaus Wallmann et al. (2018) found the hydrate release is due to the rebound of the sea bed after the ice melted. The methane dissociation began around 8,000 years ago when the land began to rise faster than the sea level, and the water as a result started to get shallower with less hydrostatic pressure. This dissociation therefore was a result of the uplift of the sea bed rather than anthropogenic warming. The amount of methane released by the hydrate dissociation was small. They found the methane seeps originate not from the hydrates but from deep geological gas reservoirs (seepage from these formed the hydrates originally). They concluded the hydrates acted as a dynamic seal regulating the methane emissions from the deep geological gas reservoirs and when they were dissociated 8,000 years ago, weakening the seal, this led to the higher methane release still observed today"[42]

So we find the methane release is, indeed, real. But it is associated with a **process started 8000** years ago. **It is not related to global warming/climate change.**

---

[42] Wallmann, K., Riedel, M., Hong, W.L., Patton, H., Hubbard, A., Pape, T., Hsu, C.W., Schmidt, C., Johnson, J.E., Torres, M.E. and Andreassen, K., 2018. "Gas hydrates dissociation off Svalbard induced by isostatic rebound rather than global warming." Nature Communications, 9(1), p.83.

Clifford Holliday

The Glaciers Are Melting

One of the impacts of global warming/climate change identified by the true believers is the melting of the glaciers. This change is hard to challenge. The glaciers have been in a melting (or receding) period for some time. Many photos and measurements verify this fact. What the cause of the glaciers' retreat is the issue. The following is a typical true believer's statement as to the cause and impact of this phenomenon.

"Why are glaciers melting?
Since the early 1900s, many glaciers around the world have been rapidly melting. Human activities are at the root of this phenomenon. Specifically, since the industrial revolution, carbon dioxide and other greenhouse gas emissions have raised temperatures, even higher in the poles, and as a result, glaciers are rapidly melting, calving off into the sea and retreating on land. Even if we significantly curb emissions in the coming decades, more than a third of the world's remaining glaciers will melt before the year 2100. When it comes to sea ice, **95% of the oldest and thickest ice in the Arctic is** already gone. Scientists project if emissions continue to rise

unchecked; the Arctic could be ice free in the summer as soon as the year 2040 as ocean and air temperatures continue to rise rapidly.

What are the effects of melting glaciers on sea level rise?

Melting glaciers add to rising sea levels, which in turn increases coastal erosion and elevates storm surge as warming air and ocean temperatures create more frequent and intense coastal storms like hurricanes and typhoons. Specifically, the Greenland and Antarctic ice sheets are the largest contributors of global sea level rise. Right now, the Greenland ice sheet is disappearing four times faster than in 2003 and already contributes 20% of current sea level rise.

How much and how quickly these Greenland and Antarctic ice sheets melt in the future will largely determine how much ocean levels rise in the future. If emissions continue to rise, the current rate of melting on the Greenland ice sheet is expected to double by the end of the century. Alarmingly, if all the ice on Greenland melted, it would raise global sea levels by 20 feet.

Clifford Holliday

> How do melting sea ice and glaciers affect weather patterns?
>
> Today, the Arctic is warming twice as fast as anywhere on earth, and the sea ice there is declining by more than 10% every 10 years. As this ice melts, darker patches of ocean start to emerge, eliminating the effect that previously cooled the poles, creating warmer air temperatures and in turn disrupting normal patterns of ocean circulation. Research shows the polar vortex is appearing outside of the Arctic more frequently because of changes to the jet stream, caused by a combination of warming air and ocean temperatures in the Arctic and the tropics.
>
> The glacial melt we are witnessing today in Antarctic and Greenland is changing the circulation of the Atlantic Ocean and has been linked to collapse of fisheries in the Gulf of Maine and more destructive storms and hurricanes around the planet."[43]

As can be seen, this is an excessively impassioned plea to stop sinning and thereby stop the glaciers from melting. It certainly

---

[43] *Hancock, Lorin*, "Why are the Glaciers and Sea Ice Melting?" 2019 World Wild Life Fund.

blames the melting of the glaciers on global warming (remember the average temperature rise in the last has been about 1.2 F in the air since 1950 and less than that in the seas). Just on the surface of the information, it seems 1.2 degrees are not much to cause such a massive melt. We will discuss this further in Chapter 4, but for now, understand it takes a substantial temperature difference to achieve heat transfer between air and water (ice.)

What about the Poles?

Why consider the Polar Regions separately at all? Well, my main question about the poles has always been why Santa Claus has chosen to live at the North Pole instead of at the South Pole. After all, there is a continent at the South Pole, not just ice and water like at the North Pole. I decided a long time ago it must be because Santa is a Northern Hemisphere (mostly European and North American) creation. Therefore Santa was consigned to the North Pole rather than to the faraway South Pole.

South Pole

In thinking further (and more seriously) about the poles, it is worth noting the South Pole (Antarctica) contains about 70% of the World's total supply of freshwater - and it is cold since it is all frozen. That fact makes it supremely important to understand all of that water (actually just a small part of the World's total water, since most – over 92% – is saltwater) sits on solid land as glaciers and snowpack. If it were to all melt, it would cause

Clifford Holliday

a significant rise in the oceans, as it is not now in the oceans. (As explained elsewhere, sea ice melting doesn't impact ocean levels, since it is already in the ocean.) This threat makes the status of this 'on land' ice important to understand.[44]

So what is happening to the Antarctic land ice? It seems there is much disagreement amongst people who study things like this for a living. What seems obvious is for whatever reasons, the area over the Antarctic Peninsula (goes north toward South America) or West Antarctic, is warming and the ice is tending to melt. The true believers claim global warming as the cause, of course. Others say there are not-yet understood things happening with the sea currents and winds. The rest of the story is where more disagreement comes. The other side of the continent (East Antarctica) is getting more ice. The disagreement is how much. The latest study I have found is dated in mid-2019, and it says over the last forty years, Antarctica has **been gaining ice** on a net basis.[45] It does indicate since 2014, this gain has reversed, but the **net is still completely positive**. Other, older, studies say there has been a continuing gain in East Antarctica but the net is negative.

North Pole

We will now move our attention to the North Pole, with our apologies to Santa for the intrusion. The chart that follows shows the Arctic Sea temperature over the year of 2019

---

[44] All data in this paragraph from a lesson plan in "Tracking Water from Space" VERSION © 2016 American Museum of Natural History.
[45] Claire L. Parkinson, PNAS, July 16, 2019.

and compares it to the average of 1958-2002. The straight (blue line is freezing (0° C), and the jagged line is the daily temperature for the year 2019. the smoother line is the average of the temperatures for the stated period. Of course, the daily measurements are much more 'jagged' than the averaged temperatures, but in general, there isn't much difference. This evaluation is particularly true for almost 200 of the 365 days (from day 100 to day 300.) Not much difference to be claiming a resultant 'melting,' particularly when **all the differences** are on days the **temperature remains well below freezing**.

Figure 14, Arctic Daily Temperature - 2019 Compared to 1958-2002[46]

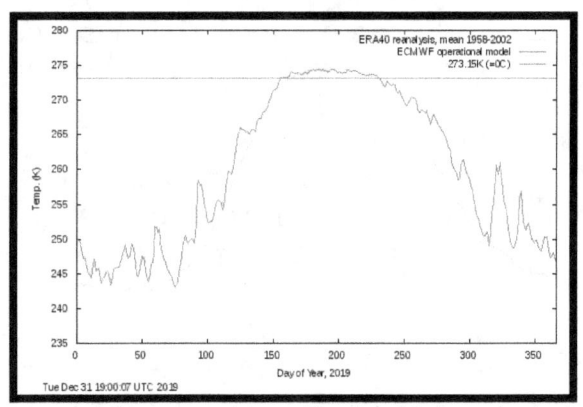

Ocean and Ice - Danish Meteorological Institute - Lyngbyvej 100 - 2100 Copenhagen Ø - Denmark.

The fact the 'difference days' were all well below the freezing point of saltwater is crucial because it shows, although the temperature was warmer, it wasn't warm

---

[46] Ocean and Ice - Danish Meteorological Institute - Lyngbyvej 100 - 2100 Copenhagen Ø - Denmark.

enough to account for the melting. <u>Other factors (other than global warming) had to have been at work.</u>

The following is a much more reasoned discussion of what may be happening to the glaciers. Its main point is **there is a series of things occurring that are not fully understood**. It makes a marked contrast to the typical true believer's assertion it is all caused by climate change.

> "Scientists already generally agree glacier retreat in Antarctica is largely being driven by warm water seeping underneath the ice—the process has been demonstrated by multiple studies in the last few years. As the ice melts, the point where it attaches to the bedrock at the bottom of the ocean (commonly known as the "grounding line") recedes inland, which can cause the glaciers to become less stable and lose more ice over time.
>
> …In an interview with E&E News about the research last week, lead study author Hannes Konrad of the University of Leeds noted the process likely has less to do with the gradual climate-driven warming of the ocean than with specific physical processes that drive naturally occurring warm water to the ice front.
>
> …Increasingly, scientists believe the winds around Antarctica are a

big part of the answer. These winds can help drive the flow of naturally occurring warm water around the continent and, in the right conditions, push it closer to the ice sheet.

Much of the warm water affecting the Antarctic ice sheet is believed to belong to a large, naturally occurring warm mass known as "circumpolar deep water." Originally formed from the mixing of waters originating in other, warmer parts of the globe, circumpolar deep water is now a fixture in the Southern Ocean."[47]

So, we see, yes, the glaciers are in a retreat period. They have done this before in cycles and are indeed melting. Yes, to the extent the glaciers melting are on land, the melt will result in a rise in the levels of the oceans. Yes, it is also apparent the North Pole is experiencing some loss of sea ice and the South Pole has some loss of ice (but also some gains of ice.) What is not explained is the connection of this melting to climate change. As the last learned quote above suggests, **it may well be a part of natural processes we just don't understand, rather than any connection to rising levels of CO2.**

---

[47] Harvey, Chelsea, "Why Are the Glaciers Melting from the Bottom: It's Complicated." April 12, 2018, E&E News.

Clifford Holliday

Increasing Wildfires

In this alleged result, the growing number of wildfires (as have plagued California and Australia most of this year – 2019 and part of 2020) is the result of global warming/climate change. This alleged result of climate change has always puzzled me. I just can't think anybody believes the small (1.75° F) change in average temperature is enough to start forest fires. However, as the reader will see in the following quotations from true believers, this is precisely what is claimed (or at least part of it.)

"Through Aug. 13, 2018, about 5.7 million acres have gone up in flames, an area larger than New Jersey. The amount burned so far this year has already surpassed the total for 2016. It's all consistent with a trend going back to at least the early 1980s: The amount of acreage consumed by fire is growing. Fires are getting bigger, and fire seasons are lasting longer. And as the planet gets hotter and parts of it get drier, it's likely these trends will continue for the foreseeable future."[48]

"There are a number of factors driving these trends. Climate change is a big one. The West, in particular, is getting warmer and drier, making it

---

[48] Ingraham, Christopher, "Wildfires Have Gotten Bigger in the Last Few Years and the Trend Is Likely to Continue," 2019, Washington Post.

easier for fires to start and spread. A 2016 Columbia University study found average temperatures in Western forests have increased by about 2.5°F since 1970, which has led to the burning of about 16,000 more square miles than would have occurred had temperatures remained the same."

"Climate is running the show in terms of what burns," study co-author Park Williams said in a statement. "We should be getting ready for bigger fire years than those familiar to previous generations."[49]

"Rising temperatures, a key indicator of climate change, evaporate more moisture from the ground, drying out the soil, and making vegetation more flammable. ...We have the power to break the cycle and get on track toward a more sustainable future."[50]

Although, in the case of wildfires, the true believers are not quite so positive in their condemnation of global warming/climate change as the cause, the accusation is still there.

Remember again we are talking about maybe 1.75° F degree increase in average temperature (in the last 70 years.) Here is a

---

[49] ibid
[50] , "Western Wildfires," 2016, Climate Central, EDF.

much more reasoned approach to assigning blame for the admittedly increasing wildfires in the US.

"In recent decades the number, severity and overall size of wildfires has increased across much of the U.S. In fact, the 2018 wildfire season in California recorded the largest fire in acres burned, most destructive fire in property loss and deadliest fires in the state's history.

But for many USDA Forest Service employees, fire season is something they remember from the start of their careers, when they quickly learned there were five seasons: winter, spring, summer, fall, and fire season. However, wildfire is year-round for much of the United States and the Forest Service is shifting to the concept of a fire year.

Wildfire season has become longer based on conditions allow fires to start and to burn - winter snows are melting earlier and rain is coming later in the fall. What was once a four-month fire season now lasts six to eight months. For example, fires in recent years have burned well outside of the typical fire season throughout California, Arizona, New Mexico, Tennessee and New Jersey. Fires in the winter months are becoming part of the norm.

Exploding the Myths of Climate Change

Other factors contributing to longer fire seasons include extended drought, <u>tree mortality from pine beetles and invasive species such as cheat grass that allow fire to ignite easily and spread rapidly. Added to all this were policies that encouraged aggressive fire suppression for more than a century. These policies had the effect of allowing fuels to accumulate, leading fires to grow in size and intensity.</u>

All these conditions are making wildfires harder to control and allowing forests to hold fire longer. For years, agencies relied on seasonal firefighters for summer months, but now wildfires are burning into the winter, they need to reevaluate their hiring plans. Wildland firefighting agencies also need to evaluate the way they conduct training for year-round fire, as well as how to handle the inevitable workforce fatigue. Residents who live in fire-prone areas must also plan and live in fire adapted communities. Defensible space, structure hardening and family plans for a possible evacuation, including pets, should be part of living in the wildland-urban interface. <u>Nearly 90% of wildfires are human-caused, so preventing wildfire is important.</u>"[51]

---

[51] Schweizer, Deb, USDA Forest Service, 2019,

Clifford Holliday
(Note: Underlining added by author.)

A recent piece in my home town newspaper (The Dallas Morning News) by a Danish scientist indicates the press calling the **Australian fires 'unprecedented' is completely wrong.** He noted Australia's **burned area has declined by a third in the 20th century**. He also noted the decrease continues as this year's (2019) **burn area was only about half of the average of the last ten years.** He further points out in his article the current efforts at CO2 reduction **are pointless, expensive, and will never achieve the desired ends of temperature reduction**.[52]

As a small aside, just a little later in 2020 from the time California was having a heat wave (alleged to be caused by global warming); Florida was experiencing an unprecedented cold snap. Freezes were experienced much further south in the peninsula than is normal. It is odd we did not read of 'global cooling' striking Florida, like we did of global warming striking California with its heat waves.

How unique that would be – global warming in California and simultaneously global cooling in Florida! **Of course, it was neither – it was weather!**

So when we look past the tiny increase in temperature blamed on global warming, we find the real experts are blaming many other things, including the ways we have approached forest management and fire suppression. In the

---

"Wildfires in All Seasons?" USDA.
[52] Lomborg, Bjorn, "Address Climate Change with Innovation," Dallas Morning News, 2-23-2020.

end, the last sentence in the last quote is most telling, "Nearly 90% of wildfires are human-caused..." This means **it was not climate change caused.**

Species Extinction

This alleged result of global warming/climate change is usually associated as a secondary impact. In other words, melting ice and rising seas cause there to be less habitat for polar bears, so they are facing extinction – except they aren't. A similar result is said to be caused by rising seas on small islands, and from a loss of grazing lands due to droughts. We have looked at many of these 'primary impacts' and found they are not getting worse, so it follows they are not causing more rapid species extinctions. Here is an example of the dire predictions.

> "A large fraction of both terrestrial and freshwater species faces increased extinction risk under projected climate change during and beyond the 21st century, especially as climate change interacts with other stressors, such as habitat modification, over-exploitation, pollution, and invasive species."[53]

Here is another dire warning.

---

[53] Field, Christopher B., editor. Climate change 2014: impacts, adaptation, and vulnerability. ISBN 9781107641655. OCLC 928427060

> "Climate change is recognised as one of the biggest threats to our natural world and its biodiversity, as well as to global security, human health and well-being.
>
> The evidence shows the cause of this change is the emission of greenhouse gases (such as carbon dioxide) into the Earth's atmosphere as a direct result of human activities, which include burning fossil fuels for energy, transport and a host of other purposes, and clearing forests and other ecosystems that absorb carbon from the atmosphere.
>
> We are already witnessing the early effects of climate change, with more frequent extreme weather events (such as droughts, flooding and storms) and changing seasonal patterns being seen around the world."[54]

The polar bear is an interesting example. Here is a recent quote from the World Wildlife Forum (in 2019.)

> "Because of ongoing and potential loss of their sea ice habitat resulting from climate change, polar bears were listed as a threatened species in the

---

[54] "A Crisis for Humanity and Diversity," 2019, Fauna & Flora International.

US under the Endangered Species Act in May 2008.

The survival and the protection of the polar bear habitat are urgent issues for WWF."[55]

Note this comment was made in 2019, even though it is well known (and has been for many years) the number of polar bears has been increasing dramatically in recent years. **Yes, in the time of increasing CO2, loss of sea ice, less food, blah, blah, blah! The number of polar bears has gone from around 5000 in the 1970s to about 25,000-30,000 now.**[56]

There is another engrossing story about the first claimed invertebrate made extinct (since it turns out it will not be the polar bear) by manmade climate change. This extinction was supposed to be the white lemuroid possum in north Queensland, Australia. However, a year later, a disappointedly healthy, thriving colony of white lemuroid possums was found 100 km south of there. Oops - not extinct.[57]

Another such instance of the first alleged species extinct due to manmade climate change was reported in 2016, claiming

---

[55] Kruger, Elisabeth, Senior Program Officer, 2019, Arctic Wildlife, World Wildlife Fund.

[56] Ibid.

[57] Nowak, R, "Rumours of possum's death were greatly exaggerated," March 2009, New Scientist.

[58] Smith, L., "Extinct: Bramble Cay Melomys". 2016, Australian Geographic. Retrieved 2016-06-17.

the Bramble Cay melomys, which lived on a tiny Barrier Reef island (Bramble Cay) north of Australia, were extinct. They were supposed to be extinct due to man-made climate change (even though they had been noted to be in decline on that island for many years.)[58] Again, this candidate to be the first victim of manmade climate change was later suspected to be thriving on Papa New Guinea.

So at this point, **we still don't have the first species to be the victim of climate change** – in spite of several false claims, all we have are dramatic estimates of pending doom.

Climate Risk

As I neared the finish of this book, I saw a report from McKinsey Global Institute (MGI)[59] I felt necessary to discuss. MGI is the business and economic research arm of McKinsey & Company, the well known and highly respected consulting company. This particular report, "Climate Risk and Response," focus on the probabilities of risks to various physical assets from climate change. It looks at risk to such things as 'Livability and Workability,' 'Food Systems,' 'Physical Assets,' 'Infrastructure, and 'Natural Capital.' The report considers possible impacts from climate change to each of these asset classes in various countries of the World.

As usual with McKinsey products, this is an extremely professional report. It leans heavily on statistical/probabilistic analysis of

---

[59] "Climate Risk and Response – Physical Hazards of Socioeconomic Impacts," McKinsey Global Institute, January, 2020.

Exploding the Myths of Climate Change

business information as is so common in expert business consulting work. I don't wish to take any issue with this report itself. My objection is how it will be perceived and used. I feel the report will be seen and taken as an endorsement of the true believers' global warming/climate change myths. Whether the authors intended that endorsement or not, I can't say. It is just obvious it will be taken and used that way.

The report itself, points out it is assuming worst case (RCP 8.5 – Representative Concentration Pathway – from the Intergovernmental Panel on Climate Change – IPCC) of anthropological climate change. This case is projected to result from taking no CO_2 abatement activities to 2030 and to 2050. The report points out it choose this RCP rather than one of the much less radical (in terms of CO_2 growth) ones in order to fully estimate the physical risk.

As long as all the readers remember the presentation is based on a worst case scenario and other unstated assumptions, there would be no misunderstanding. However, it is hard to remember these bases when you are surrounded by probability and gorgeous data representations.

The other unstated assumptions are also important. The main one is global warming is allegedly a response to increasing CO_2. There can be little doubt there is some impact from the greenhouse effect (see Chapter 1), but to pretend we understand it quantitatively is nothing but arrogance. (Again, this relationship – global warming to CO_2 – is only assumed in the report, not addressed.) <u>As we discuss in Chapter 5 on correlation, things that should</u>

Clifford Holliday
<u>correlate in relation to the global warming/climate change myth, don't.</u>

One other major assumption stated in the report is utterly weak. A couple of excerpts from the opening of the report section 'In Brief' illustrate this assumption:

> "...the Earth's climate is changing. As average temperatures rise, acute hazards such as heat waves and floods grow in frequency and severity, and chronic hazards, such as drought and rising sea levels, intensify."[60]

And the following, also from the 'In Brief' section:

> "Climate change is already having substantial physical impacts at a local level in regions across the world; the affected regions will continue to grow in number and size."[61]

The problem with these comments is they appear to be in error. Our extensive review of the data associated with the history of droughts, floods, rising sea levels and many other disasters found no relationship to the rising levels of CO_2. In fact, we found a lot of opposite indications – with many of the worst disasters in various categories taking place decades or centuries ago rather than recently as would be expected with a positive relationship to CO_2 increase. The evidence

---

[60] ibid
[61] ibid

Exploding the Myths of Climate Change

presented in Chapter 8 as relates to the various categories of disasters shows the Earth is doing embarrassingly (to the climate change fanatics) well. This pleasant state of affairs is in the face of a continually rising level of CO2.

So the bottom line on this report is it is an extremely well done piece of work (with a couple of serious questions about some of its assumptions – stated and unstated.) However, in reading it, one must remember it is an 'if this; then that' type of report. As a matter of fact, a lot of global warming/climate change writing is of that nature. If this happens; then you will get that result. The trick they like to follow with is to forget the whole thing is based on a supposition and instead treat it as if the supposition were real.

The following is a quote illustrating this type of duplicity:

> "In its Fourth Assessment Report (AR4), the Intergovernmental Panel on Climate Change warns of the dire future the Earth will face if global warming continues at the predicted rate. As many as 70 percent of extant species may become extinct if temperatures increase by more than 3 degrees Celsius per year. Millions of people may die from floods, droughts, blizzards and other weird weather patterns. Currently arable land will become arid desert, and water resources will become strained."

Nobody – not even the most ardent climate change advocate – has ever suggested temperatures may increase by 3°C in a year! Temperatures have only increased by about .5° to .6°C in the last 70 years! But here this outrageous supposition is presented as if it might be real and then continued in the dialogue as if it were. Here is duplicity of the highest order. I want to note the referenced author was not advocating this ridiculous position. He was only reporting what had been said.

So read and enjoy the McKinsey report. Marvel at it as a shining example of business consulting work and a solid warning of a worst case scenario. **But don't take it as a statement of fact!**

Summary of Effects

We have now reviewed ten (or eleven if you count famine and droughts separately) different bad results we are supposed to see from global warming/climate change. The best way to summarize the findings is **there aren't any.**

- **Yes, the atmosphere is warming – a little.**
- **Yes, the oceans are warming – even less.**
- **Yes, the oceans continue to very slowly rise.**
- **No! Flooding is not worse or more frequent!**
- **No! Droughts are not worse!**
- **No! Famines are not worse!**

Exploding the Myths of Climate Change

- **No! There are not more hurricanes and they are not stronger!**
- **No! There are not more tornados; there are actually less!**
- **No! There are not more disasters (climate or otherwise) causing more refugees!**
- **No! Insect born diseases are not worse!**
- **No the seas are not becoming acid! They are becoming slightly less basic.**
- **No! The glaciers are not melting due to higher temperatures. The noted melting is unexplained.**
- **No! The terrible wildfires are not due to the tiny temperature increase! Experts attribute them to many causes – mainly arson as in Australia recently (2019.)**
- **No! There is not a wholesale extinction of animal species due to climate change in spite of a lot of misleading advertisements. We are still waiting for the first modern species to be made extinct by climate change.**

2020???

"OK, Clif, so you have it pretty well shut down for the time frames you have gone over, but what about 2020? Surely all the fires, storms and hurricanes, and generally record bad weather have disproved your worthless theories?"

"Well, my friend, true believer, there is some truth in what you say. First, you are right, 2020 has been a horrible year, but it was bad mostly because of the Covid-19 virus, not because of climate change. You are also right we had some outlandish <u>weather</u> in 2020, but that has nothing to do with climate change, either. The year 2020 was clearly an outlier in the weather. However, it doesn't represent the latest of a continually deteriorating situation over many years, as would be expected if it were the result of, more or less, continually growing CO2."

Let's look at some of the bad weather events that happened in 2020.

Fires

First, there are terrible fires on the west coast of the US, primarily in California. These have routinely been attributed by the popular media to climate change as either causation or exacerbating fires. (See heading above 'Increasing Wildfires' for references.) The two mechanisms of climate change causing the fires have been suggested to be added temperature and drought. So, let's examine these in California in 2020. The year 2020 is undoubtedly going to be one of the hottest years on record (all of the data is not in as of this writing, but it looks like it will be in the top three, probably third behind 2019 and the hottest, 2016.)[62] Also, in terms of heatwaves, it

---

[62] Press Release Number: 02122020. 2 December 2020. World

Exploding the Myths of Climate Change
has been a scorching year for California to record highs in various places.[63] Are these two related?

The global warming anomaly is a measure of the alleged heating from CO_2 as described by global warming scientists as a difference versus a base period (1950). The second is simply one measure of the daily weather. The global warming anomaly will be about 1.75°F for 2020 **compared to 1950 – only a tiny fraction of that already small number can be assigned to 2020.** (Again, all the data isn't yet in, but this is the current estimate). Some of the daily temperature records hit 130°F as in Death Valley and were routinely 115°F at many points. So the global warming fanatics are blaming a heatwave that produced triple-digit temperatures in numerous places in California on a tiny fraction of a degree increase from the CO_2 heating effects.

Taking this one step further, the kindling temperature of dry wood in an open environment is about 500+°F [64] with many variables but all of them needing even hotter temperatures. That means to get 'spontaneous' combustion of wood in a forest (such as in California) to burn, one needs **temperatures at least above 500°F.** Of course, this is far above the heatwave triple-digit temperatures and exceptionally far above the tiny fraction of a degree of heating from CO_2 global warming.

Meteorological Organization of the United Nations.

[63] California Annual Temperatures and Records. National Climatic Data Center
[64] "Properties of Wood," Puuinfo Ltd. Snellmaninkatu 13, Helsinki, Finland.

Clifford Holliday

The heat waves did not cause the forest fires, and the little bit of temperature increase from global warming was totally insignificant.

So, what caused the fires?

Again referencing back to the above heading 'Increasing Wildfires', there is an excellent reference that spells out about 90% of wildfires are human-caused (not from CO_2 release, but from arson, campfires, inadvertent sparks, etc.) The rest are primarily caused by lightning.

Here is a reference to lightning and humans as causes.

"Even if the conditions are right for a wildfire, you still need something or someone to ignite it. Sometimes the trigger is nature, like the unusual lightning strikes that set off the LNU Lightning Complex fires in August. More often than not, humans are responsible, said Nina S. Oakley, a research scientist at the Center for Western Weather and Water Extremes at the Scripps Institution of Oceanography, University of California, San Diego.

Many deadly fires have been started by downed power lines. The 2018 Carr Fire, the state's sixth-largest on record, began when a truck blew out its tire and its rim scraped the pavement, sending out sparks. And some are started through bad decisions, like the fire that was ignited over the weekend by smoke-generating fireworks as part of a gender-reveal party and has consumed thousands of acres east of Los Angeles." [65]

---

[65] Kendra Pierre-Louis and John Schwartz, Dec 3, 2020, New York Times.

This quote also brings up the other alleged causation factor of wildfires resulting from global warming – drought. It is alleged the warmer temperatures from global warming cause the grass and brush to dry out faster and more, making it more fire-prone. (See heading above 'Drought, Famine' for a discussion of worldwide droughts and many references on the subject.) In thinking about the allegation global warming causes extreme dryness, again, we must remind ourselves we are talking about a tiny fraction of a degree in any given year. As we all know (especially our local weathermen and women), the weather is a tough thing to predict, and sometimes it brings heat and occasionally cold. When it brings warmth, it is much more significant than the scant part of a degree attributable to global warming, yet it is that little bit the true believers would have us accept as causing droughts.

A second thought on droughts being caused by global warming is to note some of the largest fires were caused by lightning coming with rainstorms. Just as a note of interest, warmer air carries more moisture, and in itself, tends to prevent fires.

Hurricanes

This year, 2020, happens to be the record year for both named hurricanes (13, through early December) and storms (30 through the same period.) So, yes, 2020 is a terrible year for hurricanes and typhoons, with many creating a large amount of suffering, property loss, and misery. The true believers would have you believe this increase in hurricane activity is due to the oceans' warming from climate change. (See the quotes and references in the heading 'Increased Hurricanes' above.)

What the global warming people maintain is these storms are worsening (they aren't in any real pattern – see the reference above). The added turbulence and intensity are due to the increase in the oceans' temperature due to global warming. The seas are claimed to have risen in temperature by .80° F since 1980 (again see references in the above chapters.) That's .8 degrees in 40 years! About an average of .05 degrees per year. (We will talk about this claim in some of the coming paragraphs, but for now, let's take it as it is.) We are asked to believe this tiny difference is making a change in hurricane patterns, even when we can't find a design. Of course, it is correctly argued the ocean is a big thing, and it holds a lot of heat. Even small differences make a big difference in the amount of heat available. All true, but, .05 degrees a year! Who believes that is causing g hurricanes?

Let's consider the number for a minute. The true believers want you to believe they can measure with enough preciseness the ocean's temperature to come up with an average for the entire ocean that is accurate down to .05° F! Think about the difficulty of measuring an average ocean temperature all over the world. Just the size and distance issues are daunting. Remember, we will be required to be accurate down to .05° F! Further, consider, you are being asked to believe there were measurement capabilities 40 years ago to produce readings compared to today's readings down to .05° F.

As I mentioned before, I am an engineer by training and practice, and I am reminded of a class I took in machine shop. Our grizzled old instructor asked us (all at least junior class engineering college students) to measure a few inches long piece of steel as accurately as we could as teams. Then the teams compared their measurements. There were surprising variations, quickly amounting to 1%, and no two groups were the same. He did this with classes every semester with the same results. That professor would laugh his head off at these claims of the accuracy of measuring something like the temperature of the oceans.

The point of that class demonstration was to illustrate all measurements come with +/- factors, even simple ones like the piece of iron. The people measuring ocean temperatures can't know the actual values with the preciseness claimed. **They can't even know whether the difference is positive or negative.**

To base a claim on the cause of hurricanes on this kind of evidence is, at best, deceptive.

The logic chain is broken!

Figure 15, Broken Logic Chain

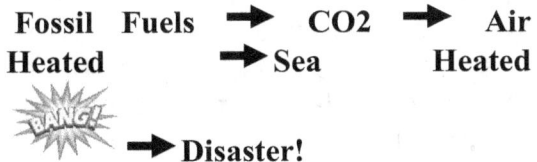

Chapter 4: Can Air Really Heat that Much Water?[66]

Myth: Hotter air is heating the seas, causing many disasters.

Truth: The physics of Thermodynamics does not support that theory. There is too much water and too little air.

So far, we have not been too technical. Warning: the material in this chapter is a little techie – but hey! I'm an engineer, and techie stuff is what we do. Seriously, the content is a bit on the technical side; it is all based on thermodynamics laws. That sounds pretty technical doesn't it? However, what's in this chapter is definitely important because it proves beyond a doubt **climate change, driven by global warming as espoused by the true believers, can't happen.**

I want to repeat what I said at the beginning of this book. I am not, in any way, contesting the facts the Earth has warmed a little; the seas continue to rise (as they have for centuries); our climate is changing (it always is.) When I say this material proves climate change can't happen, I am talking about anthropogenic

---

[66] Ideas in this chapter were originally expressed by Dr. Mark Imisides, an industrial chemist. I have not used his numbers, or his references, but have rather recreated my own to verify what Dr. Imisides has done. The reference is: Imisides, Mark, "Carbon Dioxide Can't Cause Global Warming," 2017, Principa Scientific, February.

Exploding the Myths of Climate Change

(driven by man) climate change - specifically, climate change caused by the introduction of greenhouse gases (primarily CO2) into the atmosphere, as described by the logic chain in Chapter 1 of this book.

To begin with, we will need a few quantities defined. To make you comfortable, these are almost all estimates. For example, one of them is the volume of the oceans. Well, we have many smart people who study the oceans their whole lives (oceanographers), but nobody knows what the oceans' volume is. We aren't even sure we know where all the deepest parts are – the seas are too vast to completely comprehend. But we have estimates these learned people have developed over the years, and their evaluations will do fine for our purposes. As you will see, it won't make much difference, even if these estimates are off by a factor of 10.

The first number we are will use then is the volume of the oceans. That is:

V (oceans) = 320 X 10^6 cubic miles = 1.35 x10^9 cubic km = 1.35 X10^21 liters[67]
1 cubic km = 10^12 liters

(Note: 10^12 means 10 to the 12th power, i.e., 1 with 12 zeros.)

Specific Heat of Seawater = 3850 J/kg °C[68] **= 1 Cal/gram °C**

(Note: The units in this equation may not be familiar to normal people. The first term means Joules per kilogram per degree

---

[67] Radcliffe, David, Mathbag, 2013, Wordpress.
[68] Talley, Lynne, Sio 210: Properties of Seawater, 2000.

centigrade to engineers. The next term means calories per gram per degree centigrade.)

The specific heat refers to the amount of heat in calories needed to increase the temperature of 1 gram of a substance by 1° C. [69]

Specific Heat of air = 993 J/kg°C[70]

Ratio: Specific Heat of seawater to air is = 3850/993 = 3.877

Volume (air) = 4.2 x 10^9 cubic km[71]
Mass (air) = 5.15 x 10^18 kg[72]

Density of Seawater = 1000 kg/m^3[73]

Density of Air = 1.275 kg/m^3[74]

Ratio: Density of seawater to air = 1000/1.275 = 784.3

So putting these numbers together, we can see for every kilogram of air we have 784 kg of water. We have one kilogram of air trying to heat 784 kilograms of water. Remember, from the logic chain, we are told greenhouse gases cause the trapping of heat from the Sun, and this heat warms the air and, in turn, warms the sea, causing all sorts of terrible events. When we include the impact of the difference

---

[69] "The Oceans, Their Physics, Chemistry, and General Biology," 1942, New Prentice-Hall.
[70] Lancaster, Andrew, 2016, Physics.Stackexchange.
[71] Jiang, David, "Atmosphere of the Earth," Wikipedia, 2019.
[72] ibid
[73] Talley, Lynne, Sio 210: Properties of Seawater, 2000.
[74] Jiang, David, "Atmosphere of the Earth," Wikipedia, 2019.

of the specific heats of air and seawater, this becomes:

784.3 X 3.877 = 3040.7

What this means is if we were to try to heat the entire ocean by 1 degree, we would need to heat the air to 3040 degrees hotter! **That's pretty hot**.

Summary – The Broken Logic Chain

The bottom line here is the true believer's claim of ocean heating can't be valid. Yes, it appears the oceans are heating slightly from some cause, but it's not because the air is heating from added CO_2.

Figure 16, Broken Logic Chain

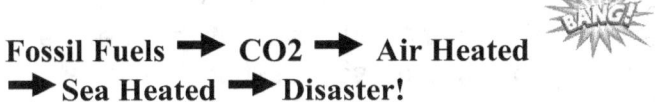

Fossil Fuels → CO_2 → Air Heated → Sea Heated → Disaster!

The logic chain is broken again!!!!

A Mind Experiment

To put this chapter in a simpler way, I have constructed the following mind experiment that goes along with the physics of this chapter.

Climate change people blame everything (heat, cold, downpours, droughts, storms, fires, etc.) on the increasing presence of carbon dioxide (CO_2) in our atmosphere. CO_2 constitutes .04% of the atmosphere. That is four-hundredths of one percent, or put another way, that is four hundred parts in a million. That amount is not much, but it can be a little hard to visualize exactly how much it is. To that point, I have put together a little mind experiment to help understand what the CC people are telling us.

Imagine a regular drinking glass of ice water (8 ozs.). Now put one drop of boiling water in it. (There are 4731 drops in 8 ounces.) How much do you think the temperature of the water would rise? Would any of the ice melt? I'll let you answer those questions, but we are not through.

Now take of ice water (with the one drop of boiling water in it) and add one more boiling water drop (actually add .89 drops.) Do you think the 8 ozs of water temperature raised any measurable amount from the two drops of water (actually 1.89 drops?) Your common sense tells you NO!

But this is precisely (or nearly exactly with a bit of rounding for illustration) what the Climate Change people expect you to believe

Exploding the Myths of Climate Change

about the impact of that .04% of CO_2 on our atmosphere. They are claiming those four hundred parts in a million of CO_2 are causing our atmosphere (the 8 ounces of ice water in the analogy) to rise in temperature noticeably and that it will cause even more increase. (The ration of the two drops to the eight ounces is the same as the .04% of CO_2 is to the total atmosphere.)

As ridiculous as that sounds, we are not finished. These people further claim that two drops of boiling water in those 8 ounces of ice water can now generate enough heat to heat another 784 times that much (784 x 8 ounces) sea water (the heating of the sea) by a degree! The '784' is the ratio of the density of sea water to the density of air.)

Those are two busy drops of water! The claim is that the oceans are being heated from the air (that was heated by the .04% of CO_2) so much that the glaciers are melting, and the sea is expanding and rising.

Still believe the Climate Change myths?

Math behind the above:
1 oz = 591.47 drops
8 oz = 8x 591.47 = 4731.76 drops
.0004x4731 = 1.8924 drops
Or
1 drop/4731.76 drops = .00021134 or .021134%

Or with two glasses 2 x .021134% = .042268% or abut the same as the CO_2 in the air

.04% = .0004 of an ounce is .0004 x 591 drops/ounce = .2362 drops

.04% of an 8 oz glass = .0004x8x591= 1.8912 drops

Chapter 5: Things Should Correlate: They Don't!

> *Myth: Global Warming/Climate Change is based on a correlation between rising CO2 and rising temperatures.*
>
> *Truth: They don't correlate.*

The idea of correlation is basically a mathematical abstraction. There I go again with all that engineering stuff. Bear with me. We will be right back to human talk shortly. Correlation is intended to express a relationship between variables that may or may not be related. It uses a 'coefficient of correlation' to give a mathematical relationship (numerically between +1 and -1) to rate how two variables change with each other. A perfect correlation is +1. At a +1 correlation, two variables move perfectly together. The associated mathematics eliminates any scale differences, so one variable may be in the millions, and the other may be in the hundreds. The issue being measured is how they move together.

It should be noted even a +1 perfect correlation is not proof of causation, but it is strongly suggestive and likely a sound basis of prediction. Likewise, a low or even negative correlation is not proof two variables are unrelated. It is, however, again, strongly suggestive.

Exploding the Myths of Climate Change

Fossil Fuels CO2 Air Heated Sea Heated Disaster!

We also use the word 'correlation' in more or less every day talk to mean merely an apparent relationship between two happenings without recognizing the more formal mathematical basis of this idea. In this sense, I want to discuss correlation, or more accurately, lack of correlation, concerning climate change. The cause for my problem here is the logic chain, repeated below:

Figure 17, Logic Chain

Fossil Fuels ➡ CO2 ➡ Air Heated ➡ Sea Heated ➡ Disaster!

In this model, which is based on all the climate change issues, the release of CO2 is the driving factor to global heating, ocean heating, and all the alleged impacts. As I have previously confessed, the increase in CO2 is an indisputable fact. The generally accepted CO2 growth looks like the following graph:

Figure 18, CO2 Over Time Graph

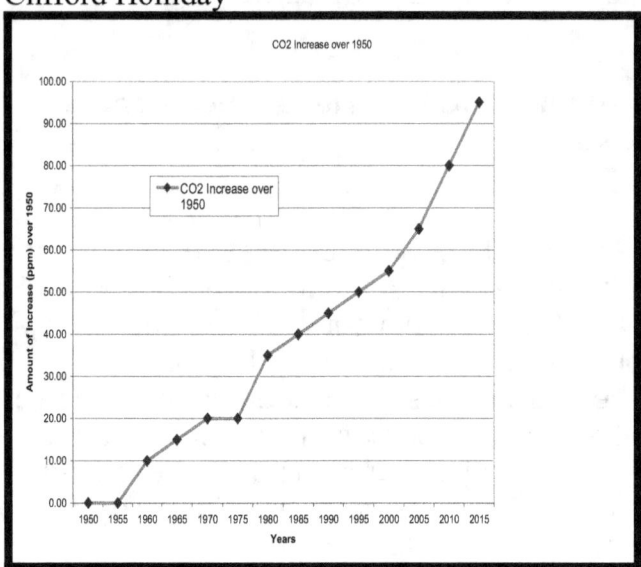

By Author – data from EPA

As can be easily seen, this graph represents a fairly continual (except for a flat few years the in the early 1970s) increasing level of CO2 in the atmosphere (even though there is still an infinitesimally small amount), presumably from its release by mankind-ordained activities. Under these circumstances and given the logic chain alleges the CO2 increase is causing global warming, one would expect every year to see hotter temperatures, more and more significant floods, more storms, warmer temperatures, etc.

The following is the customarily accepted temperature increase chart superimposed over the previous graph of CO2 to help think about this.

Figure 19, Temperature and CO2 Chart

Exploding the Myths of Climate Change

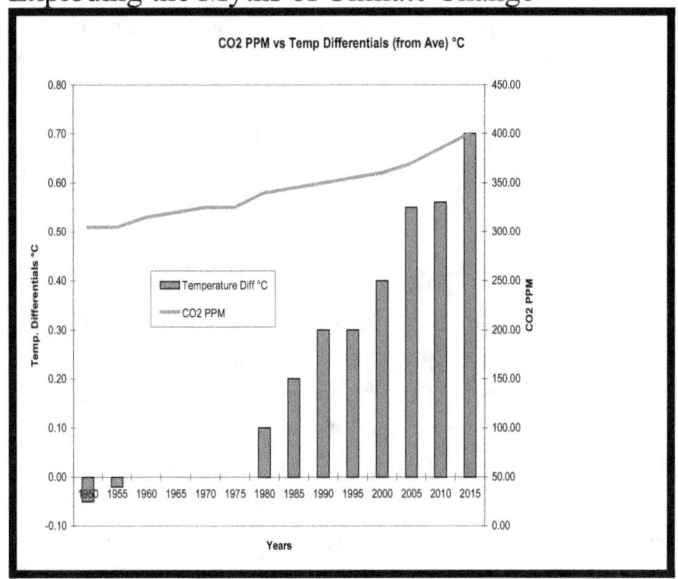

By Author – data EPA

Of course, the scales differ, so the reader is at the mercy of the graph designer because the scales can be manipulated to distort the reality. (To illustrate this possibility for distortion of viewpoints, the scales shown in the graph above make it look like the temperatures have changed considerably, while the CO2 has changed only a little. In truth, the opposite is true. The CO2 has increased by about 33% on a base of 320 ppm. The temperature has increased only about 5% on an average base temperature of 13.9° C.) However, one salient fact can't be changed. From about 1960 to about 1980, the average temperature was virtually unchanged, **even though the CO2 continued to increase.** Many climate scientists were forecasting a coming **ice age** during this time, even though it is fashionable now to deny this. (See discussion in Chapter 7.)

While the temperature was unchanged, in this time, the CO2 continued to go up at about the same rate as before. Mathematically this would be a radically low correlation (near 0) coefficient of correlation. More importantly, just from a common sense viewpoint. The two, CO2 and temperature increase, **are unrelated.**

We all hear examples like this every day on the news:

"Today was the hottest (or wettest, or driest, or whatever since 1938 (or 1890, or 1974, or some other long-ago time)"

We just don't usually think of the implication. Suppose it is the hottest (or whatever) since some time long ago. In that case, that implies a low correlation to increasing CO2 being an agent of causation (since CO2 has been growing all that time at a relatively steady rate.)

Here are some concrete examples of missing correlations, but the reader is invited to observe his/her own example almost any time one hears a report of nasty weather.

Examples:

Australia Heatwave

This country is getting a lot of press recently because of a heatwave and fires. It is indeed bad there, and many people have lost their homes, and some have lost their lives. High winds have added to the misery. The temperatures reported were a country-wide average of 105° F.

The hottest temperature <u>ever recorded</u> in Australia, however, was 123°F, **set almost 60 years ago**. At first, all the media accounts blamed global warming for the fires. Suddenly there was no mention of global warming in reports on the raging fires. It had come to light there had been multiple arrests for arson, and that was the cause of the fires. **This is not a global warming supportive correlation to rising CO2.** (All data per AccuWeather.)

Hottest Day Recorded on Earth

The hottest temperature ever recorded on Earth (indeed, there were higher temperatures in prehistoric times, but there weren't many weather stations then) was in Death Valley, California. That recording was an air temperature of 134.1°F recorded in **July 1913**.[75] (Surface temperatures can be much hotter.)
Again, not a good correlation to a steadily rising level of CO2.

India[76]

The record for this country was set relatively recently at 123.8°F. This occurred in May 2016 in Phalodi, Rajasthay. This temperature barely broke the previous high of 123.08°F set in **1886.** (All data per NDTV broadcast.)
Again, not a good correlation to a continually rising CO2 level.

Rain[77]

- Most in one minute – 1.23 inches, Unionville, Maryland. **1956.**
- Most in one hour – 12 inches, Holt, Missouri, **1947.**

---

[75] Wikipedia
[76] ibid
[77] ibid

Clifford Holliday
- Most in 24 hours – 71.9 inches, Cilaos, Reunion **1966.** (Due to tropical cyclone Denise.)
- Most in a year – 1042 inches, Cerrapunji, India **1860-61**.
- Most in a single tropical storm – 253.3 inches, Commerson, Reunion during cyclone Hyacinthe in **1980**.

The closest to today of any of these records is **1980, almost 40 years ago**. Again, **not a good correlation to worsening storms and heat being caused by constantly increasing CO2.**

For more information on the timing of various disasters, refer to Chapter 3, 'Examining the Alleged Impacts.'

Summary – Broken Logic Chain

In summary of this Chapter, we must conclude there is a notably low correlation of severe weather events to rising levels of CO2. This correlation is so small as to dispute the proposition the rising CO2 is a causation factor to weather events.

Figure 20, Broken Logic Chain

Chapter 6: "97% Of the Scientists Agree."

Myth: 97% of all scientists agree on Global Warming.

Truth: They don't

True believers can say some pretty weird things when talking about global warming/climate change. The following are ones mainly annoy me.

"97% of Scientists Agree on Global Warming."

This declaration is most often heard as a remark to end all arguments. After all, if 97% of scientists agree on anything, it must be right. As the first comment on this, I refer the reader to Chapter 7: Doomsday Predictions. Most of those were widely believed and supported by at least a large group of scientists in their time. Now, they are (almost) all considered ridiculous. Just on the face of the argument, having the agreement of a large group of scientists (or of anything) does not prove a proposition.

But we need to look specifically at this 97% assertion. After all, President Obama said it was true, so it must be the Gospel.[78] The idea there is broad agreement on climate change comes mainly from a study by Cook and others in 2016,[79] purported to review several earlier studies on how many 'scientists' actually supported anthropogenic global warming/climate change. This activity has become almost another cottage industry of its own. There are a large number of studies on this specific topic, with answers varying from 1.6% of scientists agree to 100% agree. Just the amount of disagreement makes one suspicious of the validity of the information.

500 Scientists Disagree

Several surveys come to the opposite conclusion about the opinions of the majority of scientists. Here is one appropriately recent example:

> "A global group of more than 500 knowledgeable and experienced scientists and professionals in climate and related fields representing 23 countries have written to U.N. Secretary-General

---

[78] Quoted by Alex Epstein in "'97% Of Climate Scientists Agree' Is 100% wrong," "Ninety-seven percent of scientists agree: #climate change is real, man-made and dangerous." From former President Obama's Twitter account.

[79] Oreskes, Naomi, Doran, Peter, Anderegg, William, Verheggen, Bart, Maibach, Ed, Carlton, J. Stuart, and Cook, John, 2016 "Consensus on Consensus." This was not the first time this 97% was claimed, but it is probably the most prominent occurrence.

Exploding the Myths of Climate Change

> Antonio Guterres, challenging any scientific or moral basis for prevalent alarm-based proposals."

> "Accordingly, the signatories attest it is cruel and impudent for IPCC to pointlessly and grievously advocate squandering trillions of dollars on energy programs that undermine economic systems, while denying vital access to affordable, reliable electricity."[80]

The Petition Project

Another example of a large group disagreeing with the '97%' hypothesis is the Petition Project of the Oregon Institute of Science and Medicine. Here is the petition:

> "We urge the United States government to reject the global warming agreement that was written in Kyoto, Japan in December, 1997, and any other similar proposals. The proposed limits on greenhouse gases would harm the environment, hinder the advance of science and technology, and damage the health and welfare of mankind.
> There is no convincing scientific evidence human release of carbon dioxide, methane, or other

---

[80] "500 Global Climate Scientists Challenge Mob Hysteria," October 14, 2019, Newsmax.

greenhouse gases is causing or will, in the foreseeable future, cause catastrophic heating of the Earth's atmosphere and disruption of the Earth's climate. Moreover, there is substantial scientific evidence that increases in atmospheric carbon dioxide produce many beneficial effects upon the natural plant and animal environments of the Earth.

Oregon Institute of Science and Medicine, Petition Project"[81]

These anti-global warming surveys of scientists' opinions have been heavily criticized by the true believers, but so have all the pro-global warming surveys by those with opposite opinions. There doesn't appear to be much to choose between the criticisms, except the sample sizes of the anti-global warming studies (greater than 30,000 and of course some of the replies are questionable) is so much larger than the small samples of the pro groups it deserves due consideration. My conclusion is to claim 97% of scientists agree about anything is ridiculous. To claim that an influential group agrees about global warming stretches all credulity.

In looking at one of these studies, I found what I consider a major flaw. The study questions were phrased so if I had taken the survey, **I would have likely answered in a way to be counted as a supporter of global**

---

[81] "31,000 scientists say "no convincing evidence." Open Source Systems – Science Solutions

warming/climate change, and here I am writing a book challenging that specific theory! The questions were such (Is the Earth warming – 'Yes,' of course. Are humans responsible – 'Yes,' at least partially. On their survey, those answers put me in the 'support global warming category.') As I noted at the start of this book, there is no doubt the Earth is warming – a little. There is also no doubt the greenhouse gas effect causes some of this. Besides, added CO2 from humans contributes to that – a little. It's the qualifiers that are important and the impact. As we have seen repeatedly, **it is tough to identify the harmful effects of the minor amounts of CO2 or warming. It is much easier to identify the benefits, as we will see in Chapter 8.**

Meteorologists' Survey

A recent survey of American Meteorologists had some interesting findings. These people are all practicing meteorologists, and a vast majority holds degrees (most of them advanced degrees) in meteorology. The report "A 2016 Survey of American Meteorological Society Members about Climate Change" had some fascinating conclusions. The following are a couple of excerpts:[82]

> "...say the impacts have been primarily harmful (36%) or approximately equally mixed between harmful and beneficial (36%). One out of five (21%)

---

[82] A 2016 Survey of American Meteorological Society Members about Climate Change, Center for Climate Change Communication, George Mason University, 2016.

AMS members say they don't know."

"Specifically: 29% think the change is largely or entirely due to human activity;" (Meaning 71% don't think the change is due largely or entirely to human activity.)

"…only 36 percent report primarily negative impacts from climate change in the area they cover (which, cumulatively, is the whole country)."

"…only 50 percent expect the impacts of climate change to be entirely or primarily negative during the next 50 years."

These results certainly don't support the '97% of experts agree' nonsense.

Summary of "97% agree" – <u>They don't!</u>

So, paraphrasing an old adage, "Numbers don't lie, but number manipulators can be liars." The studies present their information correctly, but it takes a detailed look to see any meaning in the numbers. **After looking through these studies of studies, my opinion agrees at least two reviewers have said, "97% of scientists don't agree about anything."** True believers may hide behind

Exploding the Myths of Climate Change
these statistics, but I think they are much too transparent to offer much cover.

Chapter 7: "What Have We got to Lose?"

> **Myth: We don't have anything to lose from following climate change and its ultimate manifestation the Green New Deal.**
>
> **Truth: We have everything to lose: our way of life; our economy; our country.**

An attitude often heard from the true believers is summarized by the thought, "What have we got to lose? If we are wrong about global warming/climate change and go ahead with our measures, we'll just leave a better world." It is curious to me this (or a similar statement) often follows an allusion to the last section in a conversation related to the following.

"Well, 97% of scientists agree on global warming. It is settled science. Anyway, what have we got to lose? We'll only make it a better planet, even if we are wrong." The curious part to me is if it is 'settled science,' why is it necessary to add the 'what have we got to lose' phrase? Are you not quite sure about 'settled science?'

What have we got to lose? Plenty!

Exploding the Myths of Climate Change

First, the cost of the true believer supported measures – the elimination of fossil fuels **would destroy several vast industries** and the **jobs associated – in the millions**. The cost of the replacement technologies (some of which don't exist) also would be enormous. Remember, the expenditure of funds for this purpose is a selection, an either/or situation. Funds spent here are not available for use elsewhere. Then there are the costs of side programs. For example, the rebuilding of every structure in the USA (as included in the 'Green New Deal'). That rebuilding would be in the trillions (and again, these are either/or funds.)

To think about this in detail, let's consider the Green New Deal as proposed in Congress (our Congress) by Representative Alexandria Ocasio-Cortez. She represents parts of the Bronx and Queens in New York City. Parts of this program have been included in President Biden's $3.7T 'Build Back Better' Bill currently being debated in the 2021 Congress. Here are the primary operational components of the Bill she introduced:

> "... (2) the goals described in subparagraphs (A) through (E) of paragraph (1) (referred to in this resolution as the "Green New Deal goals") should be accomplished through a 10-year national mobilization (referred to in this resolution as the "Green New Deal mobilization") that will require the following goals and projects—

(A) building resiliency against climate change-related disasters, such as extreme weather, including by leveraging funding and providing investments for community-defined projects and strategies;

(B) repairing and upgrading the infrastructure in the United States, including—

(i) by eliminating pollution and greenhouse gas emissions as much as technologically feasible;

(ii) by guaranteeing universal access to clean water;

(iii) by reducing the risks posed by climate impacts; and

(iv) by ensuring that any infrastructure bill considered by Congress addresses climate change;

(C) meeting 100 percent of the power demand in the United States through clean, renewable, and zero-emission energy sources, including—

(i) by dramatically expanding and upgrading renewable power sources; and

Exploding the Myths of Climate Change

(ii) by deploying new capacity;

(D) building or upgrading to energy-efficient, distributed, and "smart" power grids, and ensuring affordable access to electricity;

(E) upgrading all existing buildings in the United States and building new buildings to achieve maximum energy efficiency, water efficiency, safety, affordability, comfort, and durability, including through electrification;

(F) spurring massive growth in clean manufacturing in the United States and removing pollution and greenhouse gas emissions from manufacturing and industry as much as is technologically feasible, including by expanding renewable energy manufacturing and investing in existing manufacturing and industry;

(G) working collaboratively with farmers and ranchers in the United States to remove pollution and greenhouse gas emissions from the agricultural sector as much as is technologically feasible, including—

(i) by supporting family farming;

(ii) by investing in sustainable farming and land use practices that increase soil health; and

(iii) by building a more sustainable food system that ensures universal access to healthy food;

(H) overhauling transportation systems in the United States to remove pollution and greenhouse gas emissions from the transportation sector as much as is technologically feasible, including through investment in—

(i) zero-emission vehicle infrastructure and manufacturing;

(ii) clean, affordable, and accessible public transit; and

(iii) high-speed rail;

(I) mitigating and managing the long-term adverse health, economic, and other effects of pollution and climate change, including by providing funding for community-defined projects and strategies;

(J) removing greenhouse gases from the atmosphere and reducing pollution by restoring natural ecosystems through proven low-tech solutions that increase soil carbon storage, such as land preservation and afforestation;

(K) restoring and protecting threatened, endangered, and fragile ecosystems through locally appropriate and science-based projects that enhance biodiversity and support climate resiliency;

(L) cleaning up existing hazardous waste and abandoned sites, ensuring economic development and sustainability on those sites;

(M) identifying other emission and pollution sources and creating solutions to remove them; and

(N) promoting the international exchange of technology, expertise, products, funding, and services, with the aim of making the United States the international leader on climate action, and to help other countries achieve a Green New Deal;"[83]

---

[83] House Resolution 109, 116th Congress, Introduced in House 2/7/2019

Figure 21, Costs of Green New Deal - Estimate

Summary Table (2020-2029)		
Goal	Estimated Cost	Estimated Cost Per Household
Low-carbon Electricity Grid	$5.4 trillion	$39,000
Net Zero Emissions Transportation System	$1.3 trillion to $2.7 trillion	$9,000 to $20,000
Guaranteed Jobs	$6.8 trillion to $44.6 trillion	$49,000 to $322,000
Universal Health Care	$36 trillion	$260,000
Guaranteed Green Housing	$1.6 trillion to $4.2 trillion	$12,000 to $30,000
Food Security	$1.5 billion	$10

As the reader will note, the GND encompasses items not being discussed in this book. Items entirely outside the considerations of global warming/climate change. Of the above questions we are only interested in:

---

[84] Holtz-Eakin, Douglas, Bosch, Dan, Gitis, Ben, Goldbeck, Dan, Rossetti, Philip, "The Green New Deal: Scope, Scale, and Implications", February 25, 2019, American Action Forum.

Figure 22, Global Warming Issues of Green New Deal

Electricity Grid	$5.4 T
Zero Emissions Transportation	$1.3-2.7 T
Housing (Rebuild of Structures)	$1.6-4.2 T
Food Security (Less Meat)	$1.5 B
Total GND (Climate Change)	$8.3-12.3 T

($1.5 B lost in rounding)

There has been a lot of discussion about the estimate made above by Holtz-Eakin, Bosch, et al., commonly referred to as a "$93 trillion estimate." (The low sides of the items add to $93 T.)[85] The true believers have been suggesting it is far too high, it omits savings, and simply it is wrong. Here's one place I will agree with the true believers. Yes, like all estimates, it is wrong, and it undoubtedly excludes some offsets impossible to quantify at this stage. My problem with the true believers' stance on this is we don't hear any better estimates from them. Instead, they are only standing back, saying this is wrong. Also, there are many underestimates in the report.

My main concern with the estimate goes the other way. I think, at least, parts are probably way too low. Reading the material from these authors' reports, it seems <u>inordinately little attention</u> is given to the actual

[85] ibid.

difficulty (and cost) of converting from our existing electrical grid to a new one. Here's an excerpt from their report (previously cited):

> We estimate to transition to a power sector that has net zero emissions of greenhouse gases in 10 years would require a capital investment of $5.4 trillion by 2029. In addition, the annual operation, maintenance, and capital-recovery costs would be $387 billion. We consider this estimate to be conservative in two respects. First, we assume that a low-carbon electricity grid is feasible with only 4 hours of storage available for renewable resources; academic estimates have said a reliable grid requires 12 hours. <u>Second, we assume no new construction of transmission assets is required, even though efficiently siting new renewable assets will require significant transmission infrastructure.</u>

(Underline added by this author for emphasis.)

Note the last sentence (underlined) states the assumption of no new power line construction. As the report authors are saying, this is an unrealistic assumption and an unduly cost-minimizing one. For example, we know that here in Texas it cost approximately $10B to build transmission lines from far West Texas to the central part of the state to deliver windmill power to the population centers. These costs are still being paid off by the people of Texas. They are never considered in the calculations showing how 'cheap' wind power is. In the later section on the Texas 'Big Freeze' you will see how cheap wind power is.

The real challenge of going to the non-carbon, new grid would be siting (locating, obtaining property, access, permitting, etc.) the new power sources and then building a network to connect them to their loads. Unlike the approach envisioned by some authors and even mentioned by Elon Musk,[86] it would not be possible to locate all the new resources in a large solar (or wind, or both, and batteries) farm located in some remote desert-like part of southern California or Arizona and provision the nation from there. (It has been estimated about 21,000 square miles of solar panels could serve the entire US – see the reference just sited.) The transmission loss of the electricity from a southern California site to New York City's load would be enormous and prohibitive. The new power sites (wind, solar, and battery – everything else is minor) would have to be located all over the country, as close as possible to the main loads (big cities.) New power lines and electrical switching and protection infrastructure would be needed to connect the new power sites to their loads and new very high voltage lines interconnecting the locations to form the new power backbone grid. This construction and rearrangement would likely involve thousands of new sites countrywide, all with interconnecting power lines and support (roads, buildings, utilities, etc.) services. None of this appears to be

---

[86] Nussey, Bill, "How Much Solar Would It Take to Power the Us?" July 2108, Freeingergy.

included in the report from Holtz-Eakin, Bosch, et al.

Another thing of interest in the Holtz-Eakin, Bosch, et al. report is they included the use (extensively) of nuclear power to provide a significant portion of the electricity in states that allow it. While this, of course, makes perfect sense to assume, the same people supporting the GND are also thoroughly against nuclear power. If this GND were to be deployed, it is unlikely the people in charge would allow a significant atomic component. This inclusion would make the change much more expensive. Probably it would be more in keeping with the GND view to include the prohibitive decommissioning costs of all of the nukes.

Another series of questions come up about the housing reconstruction. This housing issue has been discussed to include rebuilding every building (residential, industrial, commercial) in the USA to be highly energy efficient. Just to get a count of how many buildings would be included is a challenging task. One estimate I have found suggests it would probably be in the neighborhood of 120,000,000 structures! The Holtz-Eakin, Bosch, et al. report appears to only cover residences, omitting commercial and industrial. Just making these additions would significantly increase the price tag. There is also a question of redundancy. If we have made the power-grid carbon-free, what is the urgency of rebuilding all the structures? I am sure the GND authors have a stupendous answer.

A Tale of Two Countries

Exploding the Myths of Climate Change

It was the best of plans; it was the worst of plans. The story of what is happening in Germany and France currently is a further tale of **what we have to lose.**

In 2020, Germany decided to initiate a grand plan to drastically reduce its carbon emissions and revolutionize its energy infrastructure. This plan, called the Energiewende, was intended to reduce greenhouse gas (mostly CO_2) by 80–95% by 2050 (relative to 1990.) It also set a renewable energy target of 60% by 2050. Somehow the Germans decided to reduce nuclear power was more important than greenhouse gases and have focused more on that aspect. (Of course, nukes don't emit greenhouse gases and have nothing to do with global warming.)

This program started in 2000 and received government support in the form of legislation in 2010. The following chart illustrates where it is as of now.

Figure 23, German Energiewende Progress[87]

---

[87] John Mathews, "The Spectacular Success of the German Energiewende," Energy Post, October 2017. The data in this section is taken from this article.

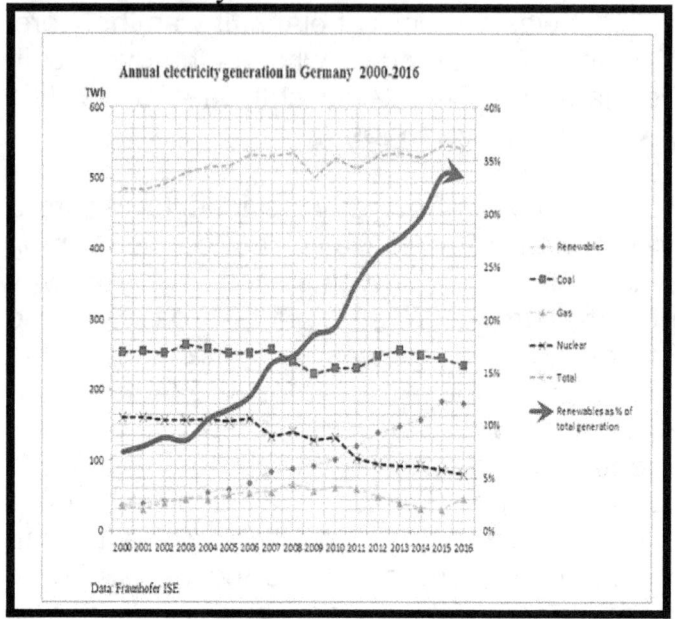

Energy Post: Fraunhofer ISE

In looking at this chart we can make several observations, most of which do not support the source article's title 'the Spectacular Success of the German Energiewende."

1st Total energy (electricity) production has increased only a little over the 16 year period. For a growing economy that is not a good sign. It means people have been deprived of all the benefits ample energy can provide.

2nd Electricity produced from coal has changed little, if at all.

3rd Electricity produced from natural gas has changed little.

4th Electricity from renewables has increased dramatically from 2-3% up to about 12%.

5th Electricity produced from nuclear plants has reduced from about 11% to about 6%.

So in summary, in 16 years the Energiewende has succeeded in stagnating the total electricity demand (at least the consumption) in Germany and has achieved a transfer from nukes to renewables. **Neither of these has anything at all to do with CO2 or global warming.**

In roughly this same time period the cost of electricity has increased by 50% for German households[88] – no wonder demand is suppressed! Also, Germany's CO2 emissions in this period have been flat – no improvement![89]

So what has this fabulous plan done? Eliminated clean, reliable nukes; made electricity enormously expensive; suppressed demand for energy; replaced the nukes with unreliable, big renewables! – <u>Good job!</u>

Meanwhile in the country just to the south, France, a different story is being told. In 1970, France began a strong conversion to nuclear energy. They have persisted in that effort to this day and now produce about 75% of their electricity through their 58 nuclear reactors. In the meantime electricity consumption has increased by 400% allowing for a vigorous economy. France has a small (16%) renewable component – largely hydro. The following chart shows the French electricity production by type in 2017.[90]

---

[88] Michael Shellenberger, "If Saving the Climate Requires Making Energy so Expensive, Why Is French Electricity Cheap?" Der Spiegel, 2018.
[89] ibid
[90] "France's Overall Energy Mix," Plante Energies, Aug 2018.

Figure 24, French Electricity Production - 2017

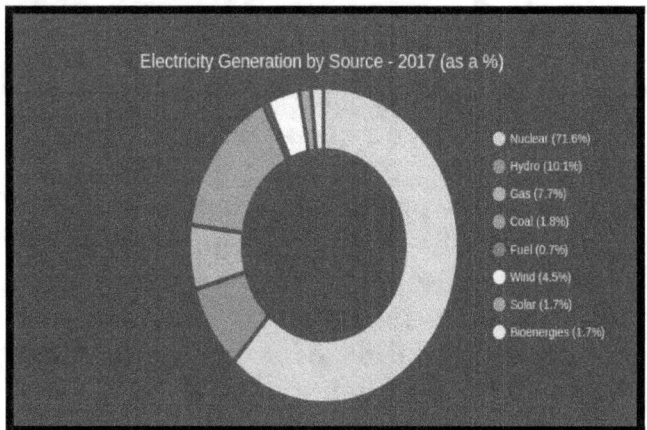

Source 2017 - Source RTE Results 2017

The result of France's program is French households pay 41%[91] less than German households for electricity and have an abundant supply. Unfortunately the story in France is changing. The true believers have taken over in the form of President Emmanuel Macron, and they are attempting to pull France toward the failing German model. Marcon has increased gasoline and diesel prices in an attempt to curb consumption. These and other

---

All the French data as well as the chart on French electricity come from this reference.
[91] Michael Shellenberger, "If Saving the Climate Requires Making Energy so Expensive, Why Is French Electricity Cheap?" Der Spiegel, 2018.

of Macron's policies are causing widespread rioting and disruption in France now. We will just have to see how this plays out. **It seems such a shame – the way of abundance and affordability is in front of them; they only have to take it. Will they?**

So What Have We got to Lose – a Lot!

So at the end of all these considerations, back to the original question, "What have we got to lose?" From the material about the cost of the 'Green New Deal' above, we have $12.3 trillion to lose in ten years (at least - as noted, this appears to be way too low). We will see later that this cost represents a choice of how we spend our funds. If we choose to buy windmills that only work sometimes, then we don't choose to buy fossil fuel plants that can be dispatched (used to add to the grid on short notice.)

That's obviously a lot of money, but more than a lot, it is so much it is hard to appreciate. To give an idea, it is about what the US Government will spend in the **next (2021-2023) three years in total!** That's for everything: debt service, Social Security, military, all social programs, everything for three years! Or, to put it another way, it is **THREE** times the cost of WW II in **today's** dollars. It's close to **FIVE** times what was collected in **all** local, state, and federal income taxes (in total) in 2020. The largest construction project in history is sometimes identified as the International Space Station. That project cost around $150 billion. (The Manhattan Project (to develop the A-bomb in the 1940s) cost $23 B in today's dollars.) That's so small it would be lost in the rounding talking about the Green New Deal!

In addition to the monetary loss (and remember dollars are just a way of keeping score - they mean real negative impacts on people), there are also additional human costs.

With that same $12.3 T, we will have **displaced over 3 million US energy workers**[92] who may or may not find jobs anyplace – almost certainly not in the area where they worked. We will have also **replaced the most reliable energy network in the world** with one highly dependent on uncontrollable occurrences (like the passing of a cloud, a prolonged period of overcast, etc). Besides, we will have **destroyed several significant industries** – oil, coal, gas, etc. We will also have **destroyed the US aviation industry** and **all the people** working in it – **some 10,000,000!**[93] What will have been done to America's farms is not certain, but it sounds ominous.

I think that is a **lot to lose, and for what**? The US is a small (maybe 20%) part of the CO_2 emission worldwide. Therefore, even if all of this effort successfully moved the US to a carbon zero status, it would **have little to no impact on the increasing CO_2** - climate change worldwide, or even on our climate.

Confusion of issues with Climate Change

It seems the young true believers, many of whom just want a better place to live,

---

[92] U.S. Energy and Employment Report, 2019, (USEER), DoE.
[93] "The Airline Industry Today," 2019, Airlines for America.

Exploding the Myths of Climate Change

confuse many environmental issues with climate change. They conflate things like smog reduction from stricter emissions standards with climate change. Smog is not a climate change issue (unless it is a cooling agent); it is an air pollution issue. Plastic in the sea, air pollution in general, water pollution all have nothing to do with climate change. Correcting problems in these areas is undoubtedly an excellent idea, **but they are not related to climate change**. As portrayed by the true believes, climate change follows the logic chain laid out in the first chapter of this book. It entails the result of increased greenhouse gases (almost entirely CO_2), causing temperature increases – nothing to do with plastic, smog, particulate emission, etc.

Chapter 8: The Texas Big Freeze

What Have We Got to Lose? – The Big Freeze showed us. The Texas electrical grid is a perfect example.

In mid-February 2021, a fearsome winter storm hit Texas. These words 'fearsome,' 'winter storm,' and 'Texas,' are seldom used together, but they applied this past February, ideally. The southern apex of what the weather people called a 'polar vortex' covered the entire state. There were record amounts of snow and plenty of ice from freezing rains. The temperatures set low records all over the state. From the northern Panhandle where winter storms are usual, 800 miles south, to the boats in Galveston Bay, where it never freezes, the entire state was covered in white.

Because it is our wonderful, ever-changing Texas (no bias here) in three weeks, those same trees had been covered with a white blanket of ice and snow were again covered in white. This time it was a blanket of flowers. Our Bradford Pears exploded into bloom with their signature white blossoms. The temperature had gone from 0 degrees to 80 degrees, and winter was over.

These highly unusual weather events led to power outages, busted pipes, hazardous roads, water main breaks, natural gas shortages, and water outages. They caused

Exploding the Myths of Climate Change

astonishing inconvenience, discomfort, and death, with at least 140 dead as a direct result!

Since the unusual weather, many cast blame in every direction, and there is plenty of blame to go around. Our Legislature has been hard at work in trying to fix the problems. Our governor now claims as a result of work completed by the 2021 Legislature that the problem is solved. One concrete change is that power plants and associated facilities must be weatherized. The Texas PUC has recently in October 2021,) passed rules requiring that insulation. We hope that the rules changes in the structure of our market plus the new insulation requirements will fix the problem. Alas, it was some of the Legislature's work about 20 years ago is primarily to blame for the problem. At the base of the proximate cause is probably the simple fact the various systems (electric power, natural gas, water) were not engineered for Texas's weather that week. The prudent businessman provides for what is likely to happen, but he does not provide for a one in a hundred years event. In Texas, what is expected to happen is sweltering weather in the summer and mild winters. Our systems are built to cope with those circumstances – not for Michigan weather in the winter. Many of the facilities – of all types: water, gas, electric – were not weatherized against the prolonged extreme cold, snow, and ice. Gas supply line valves froze; controls on all types of power supply equipment froze; windmills froze up completely or were reduced to maintenance power. Demand on natural gas supply lines was poorly prioritized, so gas-fired generators were starved for feedstock because residential gas use was given higher priority.

Clifford Holliday
Investment Choices

Sadly, Some of Texas' weather crisis was just the failure of individuals to cope with the situation. The water supply in our small city of Colleyville is a perfect example. Our mayor saw what was coming and ordered our standpipes (water towers) filled to capacity in advance. When the heavy freeze came and interrupted main feed lines to various cities' water storage, those cities lost water pressure, mold built up in the lines, and the systems had to be shut down. Not Colleyville! Our mayor had taken care of us, while all the cities around us lost water for extended periods, our water still flowed.

However, a lot of the problem was primarily because of the choices we made in power generation investments. Whether it be as private individuals, companies, or governments, we choose when we make investments. No entity has unlimited funds (although Congress this year – 2021 – doesn't act like they understand that.) When we decide to invest in "A", it is a decision not to invest in "B". Of course, we can invest lesser amounts in both, but to the extent, we invest in one, we reduce the amount we can invest in the other.

Several years ago, Texas chose to invest in renewable power sources (mainly wind with some solar.) Since we have a primarily unregulated power system, Texas had to supply incentives to get private companies to invest in wind power. Provide subsidies we did. Over the last 14 years, renewables have received over $19.4B in contributions in Texas.[94] About half of this

---

[94] Bill Peacock, Energy Alliance, "The High Cost of

came from Federal sources, but the rest came from mainly state subsidies, with a small portion from local authorities. The current plan is to provide $15.9B more over the next ten years.[95] These subsidies are in addition to the amounts paid by customers for the electricity. Subsides tend to bury the actual costs since taxes and other federal, state, and local governments' funding sources pay for the 'gifts.' However, we all know all money ultimately comes from the taxpayers and the ratepayers.

Subsidies are not, however, the 'original sin' regarding renewables in Texas. Texas's wind power origin goes back to 1999 to a bill signed into law by then-governor George W. Bush. I can't even blame my favorite whipping horse, the far-left Democrats, for this. Governor Bush signed the Texas Renewable Portfolio Standard required power companies to buy a specific portion of their power from renewable sources or pay penalties.[96] Since then, this law has developed into the significant commitment Texas has to renewable energy. Consequently, as noted above choosing the renewables investment path has resulted in curtailment of investment in non-renewables. It resulted in making the large wind farms in West Texas possible by building high-voltage transport lines and infrastructure from West Texas to the markets in Dallas-Fort Worth and Central Texas. This infrastructure consists of 3600 miles of high-voltage transmission lines

---

Renewable Energy Subsidies," quoting ERCOT and TPUC figures.
95 ibid
96 SB7 Archived 2015-09-23 at the Wayback Machine Law text Archived 2015-09-23 at the Wayback Machine Texas Legislature Online, May 1999. Retrieved September 24, 2011.

to bring power from the West and North (the Panhandle) Texas wind farms to DFW and Austin's load centers. These were built between 2009 and 2013 at a budgeted cost (ultimately to the taxpayers) of almost $6.8B.[97]

Base Load vs. Peaking

To understand the importance of these investment choices Texas has made, I need to share some information about how power grids are constructed and managed. This explanation is necessary to make my fault analysis and proposals clear. Power is provided and organized in two different chunks – base load power and peaking power. Baseload power meets the long-term, constant average requirement of the grid for power. Peaking adds power when the demand goes up based on time of day, the weather, or other events. The nature of power demand is that these peaks occur at more or less regular intervals, during the day, during a week, and over a year.[98]

'Peaking' and 'base load' are more than just terms to manage power grids. They require different types of equipment and have extremely different cost curves. Baseload generators run virtually continuously and can reliably provide large amounts of constant power over exceedingly long intervals. They are enormous, costly installations that are challenging to start and stop. Types of

---

[97] "Renewable Energy," Daniel Cusick, E&E News, February 25, 2014.
[98] EnergiMine, Jan 2021. Most of the basics are covered in this article, although there are many sources to explain the 'base' vs. 'peaking' concept.

technologies capable of this are nuclear, coal-fired boilers, and gas boiler units. All depend on heating water to convert to steam to drive steam turbines.

The following chart illustrates the baseload and peaking power concepts.

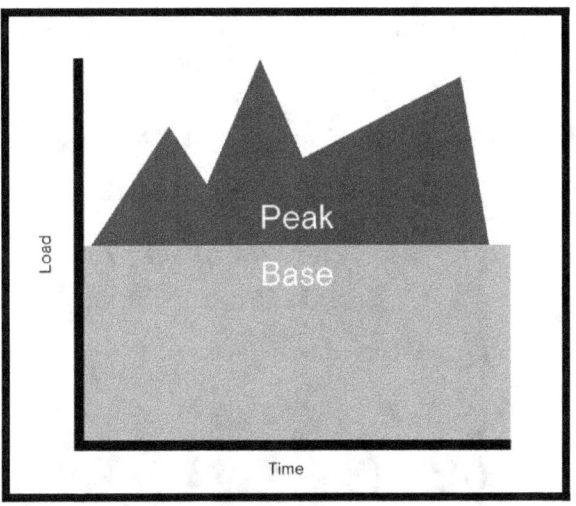

Figure 25, Peak vs. Base Load [99]

Peaking units are smaller and provide power for only shorter periods. They must be able to start quickly and reach peak power almost instantly. They are smaller and much cheaper. The type of technology most widely used for this is gas turbines, which operate much like jet engines. They meet all the criteria and are readily available.

Note the 'renewables' aren't listed in either category. This circumstance is nothing particularly against renewable energy, but windmills and solar power just don't fit either variety. They certainly don't align with the 'base

---

[99] EnergiMine, Jan 31, 2019

load' description, nor do they serve as a 'peaking' candidate. Despite the lack of accurate fit, wind power is usually included as a peaking source, and it can serve that purpose if <u>the wind is blowing</u>.

The following chart illustrates how wind energy can contribute to meeting peak requirements. Note this chart uses the terminology of 'load following' for 'peaking.'

Figure 26, Wind Energy and Peaking[100]

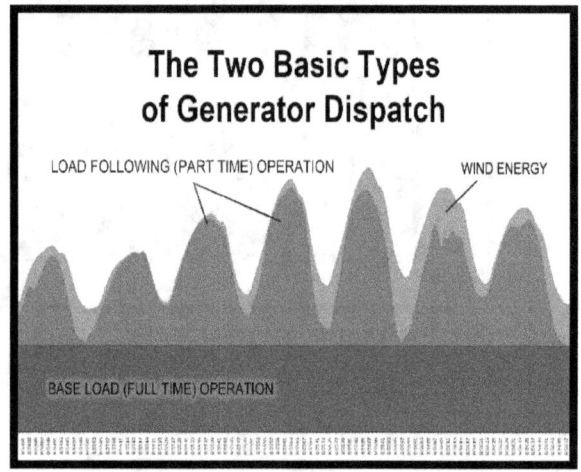

So What Really Happened to Texas' Power Grid?

With this background, let's now talk specifically about what happened to Texas' power grid in this horrible weather scenario. First, understand a few things about Texas. We have a population of over 29 million, and unlike the popular image, this population is heavily

---

[100] "FAQ – The Grid," National Wind Watch. March 2021.

centered in large cities. Texas has more cities on the top 20 in the USA list than any other state (6), and these large cities, except Dallas and Ft. Worth, are in the southern part of Texas, where it is almost always hotter.[101] As to electricity, Texas generates more than any other state and twice as much as the next state on the list – Florida. The makeup of Texas' electrical generating capacity is unquestionably diverse and has more wind-generated power than any state. Also, Texas's power network is unique because it is not connected to either of the two (Eastern and Western) national grids.[102] Texas' requirements in electricity are so large no adjacent state could be of much help.

On Feb 15th, 2021, at 12:17 AM, the first warning was issued by Texas's Electric Reliability Commission (ERCOT). About an hour later, the same commission issued orders to start rolling blackouts. The Texas power situation was unequivocally tenuous for the next few days. The blackouts continued for much longer than the usual 2-3 hours of rolling blackouts; there were boil water orders out in many jurisdictions – ultimately nearly half the state, but no electricity; water systems were shut down completely; some areas had meager gas pressure; it continued to be extremely cold. It was a terrible time for Texans.[103]

Behind this series of tragedies, the electric grid had its winter vulnerabilities

---

[101] "City and Town Population Totals 2010-2019," United States Census
[102] ERCOT, 2017
[103] "The Texas Freeze: Why the Power Grid Failed." Katherine Blunt, Wall Street Journal, February 20, 2021.

Clifford Holliday

exposed. The total capacity as published by ERCOT as:

Natural Gas 51%
Wind 24.8%
Coal 13.4%
Nuclear 04.9%
Solar 03.8%
Other 01.9%

Figure 27, Power Sources in Texas by %[104]

To calculate the totals in megawatts of power (MW), we know nukes have an installed capacity of 5000 Mw at the state's two nuclear plants.[105] So those figures imply a total of about 102,000 MW installed power in the state. Based on this calculation, the source chart above can be rewritten in terms of megawatts as follows;

---

[104] "What percentage of Texas' Energy is Renewable?" Nate Chute, Austin-American Statesman, February 17, 2021.
[105] "State of Nuclear Energy in Texas," Stateimpact, Texas, February 2021.

Exploding the Myths of Climate Change

Natural Gas 52,000 MW

Wind 25,296 MW
Coal 13,668 MW
Nuclear 5000 MW
Solar 3876 MW
Other 1938 MW

Total 101,778 MW (difference due to rounding)

Figure 28, Power Sources - Texas - MW

By 3 AM Monday (February 15th, 2021), wind generation had dropped 32%; coal 13%; and gas 25%. Besides, at about 5:30 AM that morning, one of the two nuclear units at the South Texas Nuclear Power Station dropped off-line taking out 1280 MW of capacity. (About a quarter of Texas' total atomic capacity.)[106]

During this period, many gas-fired plants dropped off-line as well as some of the coal-fired ones. In total, 185 generating units (out of approximately 710)[107] went off-line. Wednesday's total deficit was estimated at 46,000 MW off-line, with 28,000 of the deficit coming from coal, gas, and nukes and 18,000 MW coming from the wind and solar sources.[108]

---

[106] ibid
[107] "What Percentage of Texas' Energy is Renewable?" Nate Chute, Austin-American Statesman, February 17, 2021.
[108] "Why Coal and Gas Went off during the Texas Cold Weather," Abby Smith, energy and Environment, February 19, 2021.

Clifford Holliday

The following figure shows the power generation in Texas during the critical time February 11th, 2021 through February 17th, 2021, by source. Note (as has been loudly proclaimed by certain 'journalists') all sources – gas, coal, nukes, and wind – were affected by the weather disaster. Also, note gas dropped the most in absolute terms, going from an average (just reading the graph) of about 35000 Mwh to 28,000 Mwh, or about a loss of 7,000 Mwh. Coal dropped about 3,000 Mwh; nuclear about 1,000 Mwh; and wind from an average of nearly 6,000 Mwh on February 14th to almost zero on February 15th, for a loss of about 6,000 Mwh.

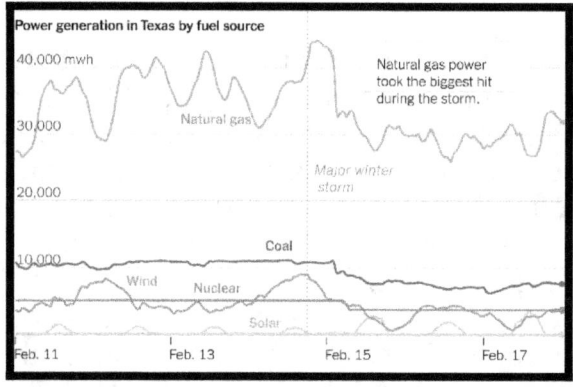

Figure 29, Texas Power Generation by Source - Critical Period[109]

At first, it was thought the gas units failed because of moisture in control valves freezing, and that is undoubtedly true to some extent. However, it has since been discovered at least a significant part of the problem was a

---

[109] By The New York Times Feb 19, 2021: Source: U.S. Energy Information Administration Hourly Electric Grid Monitor,

human failure. The gas-driven electrical generator units' operators had not applied for priority on their gas supplies and were cut off when gas pressures ran low. A form had to be filled out and submitted to the gas suppliers to be high on the priority list. The operators of many generating units had failed to do so. Since they weren't a high priority, they were cut off when gas supplies were limited.[110]

The windmills hadn't been weatherized for this kind of weather. Windmills operate in the northern climes usually well, but they are prepared for it. Texas' windmills weren't. Reports indicated half of the state's windmills had frozen, and the remaining ones were operating at a meager output. To understand the impact of the loss of wind (and of course, it was night time so solar was at zero), the following comparison is offered. Just two weeks earlier, the windmills provided power for between 40% and 50% of the total Texas load during the day and no lower than 20% at night. By the morning of Feb 15th, windmills fell to about 10% of full load and almost zero before the day was over. That loss caused a giant hole in the operator's capabilities.[111]

The nuclear unit went off-line due to a freezing feed water unit on the 'non-hot' side of the generator. It was back online at total output later in the week — all of the other nukes produced at capacity through the crisis.

---

[110] Erin Douglas, "Paperwork Failures Worsened Blackouts," Texas Tribune, as re-printed in the Dallas Morning News, March 21, 2021.
[111] ibid

Clifford Holliday
Our Sins

As noted at the beginning of this discussion, the original sin here was constructing the $6.8B worth of high lines from West Texas to give the wind farms a market. It could easily be contended spending that money to encourage more reliable generating capacity (weatherized) would have reduced the likelihood of the 'Texas Freeze' ever happening. However, that sin has been committed. It can't be undone. Now we must make the best of it. The windmills do serve a beneficial purpose in the summer. The winds blow the most in West Texas in the summer, and the windmills produce much more reliably then. They are a source of essentially free power (ignoring the substantial costs of maintenance and depreciation), as was initially envisioned. We need to ensure they are winterized so they will produce in the winter, too.

The second sin, however, continues. That is pouring subsidies into wind and solar energy. We need to quit trying to pick technology winners and instead encourage investment in all electrical power types.

Our current market structure offers little encouragement to baseload plants that produce all the time, including the majority of the time when the 'bid' price for electricity is meager. Peaking plants tend to produce only when the bids are high and make more money for the time they run. This anomaly has caused the Texas grid to become heavily slanted toward the peaking plant. We need to make some changes to encourage baseload plants' development, even though this is a long-range activity due to lengthy permitting and construction times.

What are the Wages of Our Sins?

So what has the 'wind bias' cost us? **It has cost us a reliable power system in Texas.** We have poured billions of dollars and encouraged others to spend many more billions on an unreliable grid network. We have made investment choices encouraged by subsidies rather than by market realities. We have vast capacity, but only a 'half-vast' capability to use it when we need it the most.

How do We Fix This Mess?

So how can we fix this? The sinning investments have been made. That can't be undone. However, as we advance, we do have some options. After reviewing everything that happened, the author offers the following recommendations.

Specific grid and other recommendations to fix Texas' power issues:

- First, fix the human failings. Be sure the paper work is completed to get gas-driven generators on a high priority for natural gas delivery. Be sure all cities have plans to address coming severe storms, e.g., the proper measures to be taken to assure a water supply.

- Require the weatherization of critical elements that deliver gas to our baseload gas plants and our gas peaking plants. This weatherization will primarily be of the gas delivery infrastructure and may need cross-regulatory authorization. Likely, our entire infrastructure doesn't require weatherization. (Remember, it tends to get hot in Texas, which requires the opposite kind of weatherization).
- Stop any subsidies that attempt to pick technological winners. Let's not make any more bad investment choices.
- Require weatherization of windmills if they are to continue to receive any state subsidy. Restructure the state boards that control the various systems – natural gas, electricity, and water – to at least work together.
- **Encourage the development of increased baseload generating capacity. An increase in our baseload capabilities will ease the pressure on peaking units.** Include a provision to give incentive to baseload production via assured returns for baseload. This incentive needs to be built into our open market structure. Other states have a similar feature. A baseload generating company went out of business just a couple of years ago. We can not allow that. Baseload generation must be kept profitable for our grid to remain viable.
- Require at least a percentage of any new generator construction based on weather-

dependent technologies (windmills, solar) include fully backed up and wholly weatherized sub-systems.

The first four of these need to be accomplished as much as possible by next winter (2021-22.) The rest are more extended term corrections to allow us to meet future load growth needs sensibly.

We will see if the actions taken by the 2021 Legislature and followed up in October 2021 by the PUC will correct the problems.

Texas – Summer of 2022

Sometimes you just can't win. After our terrible freeze in Texas in 2021, we are following up in 2022 with record summer heat. To add to the 100+ degrees virtually every day for many weeks, it is now about 50 days since we have had ANY rain in the Dallas – Ft. Worth area. Most of the state is under a drought warning, of course and the extremely dry conditions added to the carelessness of some of my fellow Texans have led to many grass fires. Naturally, the Climate Change fanatics are loudly proclaiming that this is all proof of global warming. The point that they seem to miss (actually one of many, but who's counting) is that most of the heat records that are being broken are not all time records, but typically go back many decades, some even to the 1800s, at which time it was hotter. Not only was it hotter at some long ago past time, that past time was almost always long before the buildup in CO_2 occurred. Whoops! What happened to the 'greenhouse gas' theory?

Clifford Holliday
The Grid in 2022

So how has the grid done in the stress of the extreme heat of the 2022 summer? Overall it has done pretty well. We have had a few days when the power authority asked the public to conserve, but we have had no blackouts or anything like that as this is being written (July 2022.)

However, the wind has been very still during this period of high temperatures. And, as you know, when the wind doesn't blow, the windmills don't go. The contribution of our largest in the nation fleet of windmills has produced very little toward satisfying the record electricity demands this summer. During a recent period, the wind power only produced 8% of its capacity. To illustrate how much of a problem this is, I have constructed the following mind experiment.

'What if' Experiment – Texas Grid

To understand our power problem here in Texas, let's conduct a quick imaginary example. If over the past few weeks (June and July 2022), we had had 100% wind and solar power, we would have had emergencies everyday with at least rolling blackouts and maybe we would have completely lost the grid. On the other hand, if we had 100% fossil plants (instead of wind) during that time, there would have been no emergencies, no calls for conservation and never a blip in our power availability.

The calculations behind the above follow: (US DOE data -2022)

Exploding the Myths of Climate Change

Texas total installed capacity = 93,219 Megawatts (128,947 installed nameplate, summer capacity)
Fossil Fuels = 53,000 MW
Nukes = 5 MW
Renewables =38,000 MW
Wind =23,800 MW
Solar =13,800 MW

Peak demand summer of 2022 = 80,000 MW (APPROX)

Wind contribution during most stressed hour = 8% of capacity

If all fossil was wind; wind from Fossil = .08 x 53,000 = 4,240 MW
Wind existing =
.08 x 23,800 = 1,904
Solar = (assuming 100% available
1.0 x 13,800 = 13,800
Nukes (assuming 100% available)
1.0 x 5,000 = 5,000
Total
= 20,704 MW

So with the 'convert all fossil to wind' scenario we have a production of 20,704 MW and a demand of 80,000 MW result; long blackouts, and almost surely the loss of the entire grid.

Clifford Holliday

On the other hand, if it were all fossil (i.e., convert all the wind power to fossil) then the scenario would look like this:

Wind converted to fossil =
1.0 x 23,800 = 23,800 MW
Solar = (assuming 100% available
1.0 x 13,800 = 13,800
Nukes (assuming 100% available)
1.0 x 5,000 = 5,000
Fossil (existing)
1.0 x 53,000 = 53,000
　　　　　Total
　　　= 95,600 MW

Now we have a production of 95,600 MW and a demand of 80,000 MW. That is a 15,600 MW margin – the grid meets all requirements with out a murmur.

Chapter 9: Doomsday Predictions that Didn't Go 'Bang!' – The 'Cry Wolf' Syndrome

Myth: The sky is falling; the sky is falling!

Truth: It isn't; Myth makers over the centuries have been sounding false alarms.

We all know the kiddie fairy tale about the little boy who cried wolf too many times. When the real wolf came, nobody would answer his cries. I fear that has happened in the realm of climate change. Today we hear a lot of dire forecasts from true believers. These forecasts are like the doomsday forecasts we have listened to over our lifetime (and, indeed, over the ages) that didn't pan out. Making forecasts is a challenging activity, especially about the future. Predictions can alert people to what could happen. However, as we will see, forecasting, particularly **about weather/climate, quite often tells us about what will not happen.**

Still, when the forecasters take themselves too seriously and want the entire World to change its way of life, **forecasting is more than an idle pastime. It is fraught with many visible disastrous results. It is hazardous to all our health.** We are now hearing from many of the true believers outrageous claims of pending doom: "...ten years before the Earth changes irreversibly," "...seven years to the 'tipping point,'" "...eliminate fossil fuels by 2030 or we will be doomed."

To put some of the current claims into perspective, we will review some made in the last century and in antiquity. Of particular interest, of course, are those that relate to climate change. These doomsday predictions sound like the ones being made now based on the true believer's models. Are these new predictions any better than those listed? As we have shown, the actual record of what is happening **does not support the model-based doomsday predictions that are now so much of the true believers' narrative.**

Some of these forecasts of doom are presented somewhat tongue-in-cheek. However, keep in mind each was serious at its time. (This is just a sampling of such predictions. I could quickly fill this book with similar examples, but a few will make the point. It seems mankind has had a love affair with predicting the end of the World all through history. Perhaps this penchant comes from a deep dissatisfaction with the way they see the current World.) Each was highly supported by at least a sector of the then 'scientific' community. Each was believed by many – millions in most cases – of true believers. **And each was utterly WRONG!**

Al Gore

This is one that obviously must come first; those looming disaster claims made by our former Vice President (and self-proclaimed creator of the Internet) Al Gore, are the top of this pile. In 2006, Mr. Gore predicted several

Exploding the Myths of Climate Change
things in his book and movie, "An Inconvenient Truth." These predictions included;

- the oceans could rise by 20 feet;
- the Arctic Ice would disappear by 2016;
- polar bears would face extinction due to the lack of Arctic ice (see discussion of this falsehood in Chapter 3,
- we would have reached a 'tipping point,' a point of no return in climate change, by 2016.

In fact;

- the oceans rise about a 1/8 of an inch a year and have been doing so for thousands of years;
- there is less Arctic ice, but it is by no means gone;
- there are more polar bears, by at least a factor of 5 times, now, than there were 30 years ago;
- and while CO_2 has been going up rather steadily, the temperature is barely increasing and we have a rewardingly fruitful, bountiful World.[112]

AOC

Alexandria Ocasio-Cortez, the United States Representative from Queens, New York. Although her predictions don't fall into this category of 'predictions that didn't come true,' we would be remiss if we didn't mention some of her outlandish thoughts. She has predicted

---

[112] Bullitt, J. Frank. "Gore Says his Predictions Are Coming True. Can He Prove it!" August 2019, Issues and Insights.

Miami would disappear in a few years, and that wouldn't matter much because she also has predicted the World will end in 12 years if something isn't done about climate change. AOC's answer is her 'Green New Deal' (see Chapter 6 for details on the GND), endorsed unbelievably enough by several of the 2020 Democratic candidates for president's nomination.

This Green New Deal includes such things as;

- rebuilding every structure in America to make them more weather efficient;
- do away with planes (implicitly);
- do away with internal combustion cars;
- do away with current trains;
- stop producing and eating beef (that is not the greenhouse gas CO_2 they expel, AOC, it is hydrogen sulfide, H_2S, and some methane, CH_4).

In other words complete destruction of our way of life. The cost of this plan has been estimated to be as much as in the 100's of trillions of dollars[113] – more money than exists, but its cost doesn't matter, it will never happen.[114]

We should be ever vigilant however. As the saying goes 'there are many ways to skin a cat.' The Biden Administration is proposing a multi-trillion dollar 'Infrastructure Bill.' Everyone is in agreement much of our infrastructure – roads, bridges, rail, airports, etc – is in need of

---

[113] Better cost estimates are presented in the Chapter on "What Have We Got to Lose?"

[114] See Chapter on "What Have We Got to Lose?"

repair or replacement, and that is the way the bill is being sold. However, in truth, much of the expenditure of the bill is directed at starting the 'Green New Deal.' To use another saying, 'it is a wolf in sheep's clothes.'

This Bill has now been split into two 'Infrastructure' Bills. The first, which has bi-partisan support is about improving bridges, roads, the Internet, etc. – real infrastructure. The second Bill, much larger ($3.7T) is the beginning of the Green New Deal and not about infrastructure at all. It is backed strictly by the very liberal wing of the Democrats in Congress.

The Coming Ice Age!

This ice age prediction is a real shocker to some of the younger true believers, but many of us are old enough to remember the headlines first-hand. In the 1970s, many 'serious' scientists were predicting a new ice age was upon us. Here are some of the quotes

> "In the next 50 years, the fine dust that man constantly puts into the atmosphere by fossil-fuel burning could screen out so much sunlight that the average temperature could drop by six degrees. If sustained over several years, such a temperature decrease could trigger an ice age"[115]

---

[115] Rasool, Dr. S I, NASA, and Columbia University, July 9, 1971, Quoted in the Washington Post.

Clifford Holliday

> *In a letter to President Nixon in 1972, from Brown University, Department of Geological Sciences, The Department Chairman, and Dr. George J. Kula; of the Lamont-Doherty Geological Observatory, warned of "global deterioration of the climate ...cooling falls in the rank of processes that produced the last ice age ...present rate of cooling seems fast enough to bring glacial temperatures in about a century... substantially lower food production... increased frequency and amplitude of extreme weather... bring floods, snowstorms, killing frosts, etc."* (Sound familiar?)[116]

The current true believers are so embarrassed by this particular set of predictions they tend to brush them away as media hype. However, if one pursues the literature from the 1970s, <u>it is all there.</u> Many 'serious' scientists **were advocating that an ice age was more of less eminent.**

The debate about this has raised a lot of interest because, if the 1970's consensus about an ice age was wrong, then what is to say the alleged '97% consensus of current scientists' about climate change is right? Here is an excerpt attacking the ice age consensus;

> "In their 2008 paper, *The Myth of the 1970s Global Cooling Scientific Consensus*, Peterson,

---

[116] Rasool, Dr. S I, NASA, and Columbia University, July 9, 1971, Quoted in the Washington Post.

Connolley, and Fleck (from now on PCF-08) state, *"the following pervasive myth arose: there was a consensus among climate scientists of the 1970s that either global cooling or a full-fledged ice age was imminent…**A review of the climate science literature from 1965 to 1979 shows this myth to be false. The myth's basis lies in a selective misreading of the texts** both by some members of the media at the time and by some observers today. **In fact, emphasis on greenhouse warming dominated the scientific literature even then.**""*[117]

It turns out the study Peterson and Connolley based their assertions on was a minimal review of papers then current. The results of a much broader survey are indicated in the following quote:

"The 1968-1976 period when cooling papers greatly outnumber the warming papers (85% to 15%), if we ignore the neutral papers (as was done in the Cook et al. (2103). The 85% to 15% majority is an overwhelming cooling consensus. Additionally, this is probably the period when the 1970s *"global cooling consensus"* originated because cooling **was an**

---

[117] McFarlane, Angus "The 1970s Global Cooling Consensus was not a Myth," 2018, Climate News.

Clifford Holliday

established scientific consensus – not the myth that PCF-08 contend."[118]

It doesn't appear an ice age is coming, but it sure was **forecast by many 'climate scientists' in the 1970s.**

New York's West Side Highway Underwater by 2019.

This prediction was made by Jim Hansen, the scientist who also forecast the greenhouse effect. This West Side Highway underwater forecast was uttered in 1988 and reported in the press in 1989. **So far, the West Side Highway is above water and, although extremely busy at times, doing well.**[119]

No More Snow

On a recent night here in Texas, I had the memorable pleasure to see and hear the Texas Tenors sing White Christmas. However, some true believers are telling us we may as well stop dreaming of a White Christmas because there will be no more snow in just a few years. A senior research scientist in Britain (Dr. David Viner of the Climatic Research Unit of the University of East Anglia) predicted (in 2000 almost 20 years ago) the end of snow. He said snowfall will become a rare and exciting event in a few years, due to global warming. "Children will not know what snow is." [120]

---

[118] ibid
[119] "1989; New York's West Side Highway underwater by 2019," June 1989, AP.

Exploding the Myths of Climate Change

Although we don't have much snow in this part of Texas (once every five years), I guess the children in the US Midwest this year (2019) know pretty well what snow is as they have had several blizzards. Again, **a prediction that hasn't happened, and in fact, the opposite is happening.**

Nostradamus Predicted the End of the World.

A quatrain by Nostradamus, which stated the "King of Terror" would come from the sky in "1999 and seven months" was frequently interpreted as a prediction of doomsday in July 1999.[121]

It didn't happen!

The Mad Monk Predicted the End of the World

Rasputin, a Russian mystic who died in 1916, prophesied a storm would take place on August 23, 2013, where fire would destroy most life on land, and Jesus would come back to Earth to comfort those in distress.[122]

It didn't happen!

---

[120] "2000: Children won't know what snow is." September 2015, CEI.org.
[121] Randi, James, "The Mask of Nostradamus." 1993, Prometheus Books. ISBN 978-0-87975-830-1
[122] "Grigory Rasputin predicted end of the world on August 23, 2013". Pravda.ru. August 23, 2013. Archived from the original on August 7, 2017.

Clifford Holliday
Summary of 'Crying Wolf'

These examples could go on to take up more pages than this entire book. However, the point is we have been hearing horror story predictions for a long time (centuries.) What we have listened to before sounds extravagantly similar to what we hear now. While all are predicting doom, many of these stories are **opposite in causation** to what is currently being claimed. This juxtaposition has made it necessary for the true believers to try to go back and change the record of what was said by many 'climate scientists' just a few decades ago. This is the case of the 'coming ice age' claims. Now there is an attempt by the current global warming true believers to erase that record. However, the history is still there as an embarrassment to the narrative currently being sold.

We still hear cries of 'wolf,' 'wolf,' but **we owe it to ourselves and our coming generations to be sure we don't 'kill the goose that is laying the golden eggs' just because of a 'wolf' cry.**

Chapter 10: With All These Bad Effects, How Are We Doing?

Myth: Global Warming/Climate Change is causing increasing numbers of and severity of all kinds of natural disasters, and it will get worse.

Truth: It isn't! Natural disasters of all types are at low ebb and seem to be getting lower, when looked at over time.

Let's repeat the 'less common' effects from Climate change listed earlier in the book. They were:

- Flooding
- Droughts, famine
- Increased Hurricanes
- Climate refugees
- More insect born diseases
- Acidification of the seas
- Release of methane from the permafrost melting
- Glacial Melting
- Fires
- Species extinction

Clifford Holliday

We have looked in detail at all of these and found no substantial relationship between these events and anthropogenic global warming/climate change caused by increasing greenhouse gases. We have reviewed each of these and found in most cases, the worst event in the category (e.g., worst flood, drought, etc.) happened long ago, not recently, as would be the case with increasing CO2 being the causal agent. We have also found in most cases, there is an improving situation (e.g., more polar bears now than decades ago) rather than a deteriorating situation as we have **purposely** been led to believe. This lack of adverse effects, along with other issues identified herein, has led us to the **assertion anthropogenic global warming/climate change, with its attendant list of damaging effects, is a hoax.**

We have reviewed some of this hoax's potential costs in Chapter 6 entitled "True Believers Say the Damnedest Things" and found **by adopting the global warming/climate change 'fixes' espoused by the radical leftwing, we have a lot to lose.** We now want to look at the more positive side of affairs. The question now is, "How Are We Doing?" To answer, we will review some metrics about the USA's and the World's quality of life and how that quality has been changing.

Poverty Level

To start with, let's see how our poverty levels stand. The following chart from the US Census Bureau shows how the US poverty level has trended for the last few years.

Exploding the Myths of Climate Change

Although not on this chart, 2018 and 2019 also saw reductions in poverty percentages. As we look at this chart and others in this chapter, let us do so, remembering any poverty level above 'zero' is terrible, and our efforts must continue to move to that goal of 'zero.'

Figure 30, US Poverty Levels[123]

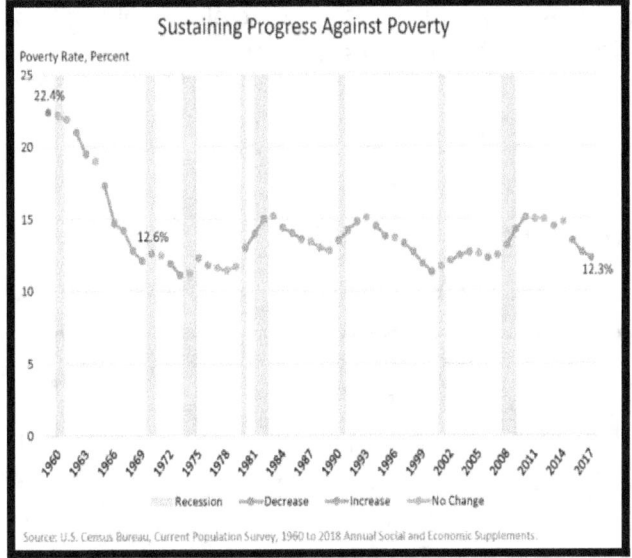

By US Census Bureau

At 12.3% for the end of 2017, the US is one of the lowest points on the chart. It is also notable during this period (which shows sustained increases in CO2); the poverty rate has decreased almost by 100%. This rate was down again in 2018 to 11.8%. It appears to be going down in 2019, but the final data is not available at the time of this writing. So the

---

[123] Edwards, Ashley, "Poverty Rate at 12.3 Percent, Down From 14.8 in 2014," September 12, 2018, US Census Bureau.

poverty rate is not only low; it is the lowest in US history and appears to be going lower, still. This has occurred in a time of increasing CO2, and allegedly (AOC and others) the coming destruction of our planet.

What about the World poverty levels in this long period of increasing 'carbon pollution'? The following chart will show much the same thing for declining World poverty. The reader will note a vastly different measure used for world poverty, but we are only looking for trends, so that won't impact our purposes.

Figure 31, World Poverty[124]

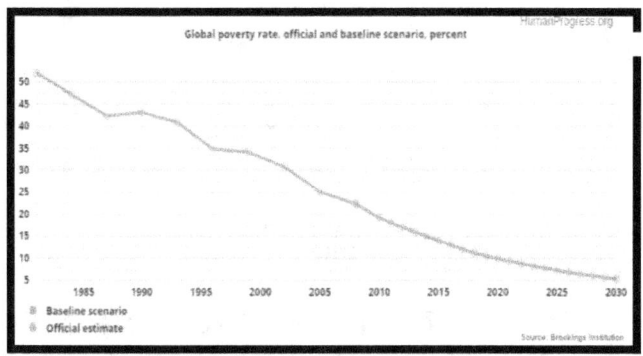

By Brookings Institute

:ars after 2017 are forecasts.

As can be seen from this figure, World poverty is also trending downward dramatically. So the bottom line is worldwide, people are

---

[124] Brandt, Tyler, The Daily Bell, "6 Reasons for Optimism in 2020"
 - December 30, 2019 – Data from the Brooking Institute

Exploding the Myths of Climate Change

enjoying the best standard of living (at least, substantially fewer are dealing with abject poverty) in this period of rising CO2.

Crop Production

One of the claims from the true believers is global warming/climate change is creating weather conditions that will be detrimental to crops and cause widespread famine. Let's examine where we are on the food curve. The following chart on crop yield per acre (corn in this case) illustrates the ability to grow large quantities of corn in the US on not much land, and how that has been dramatically improving in recent times.

Figure 32, US Corn Yield[125]

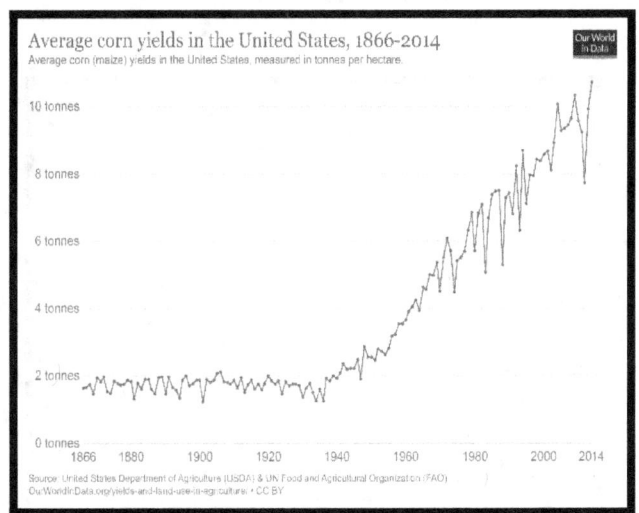

By USDA

---

[125] Ritchie, Hannah and Roser, Max, "Crop Yields," September 2019, Our World in Data.

Just a glance at this chart shows corn yield in the US has increased by almost 400% in the interval from 1940 to today. It has been mentioned before, but it needs to be emphasized: **CO_2 is vital to plants**. In fact, CO_2 is necessary for plants to live. **<u>Plants, like all life on Earth, are carbon-based</u>! CO_2 is not a pollutant; it is the basis of our form of life!**

The following chart will illustrate the same kind of agricultural productivity improvement on a global basis. Pure worldwide data is tough to obtain because of the differences in crops, etc. However, the data in the following chart from the United Nations should do nicely to illustrate the point food production is going up, worldwide, not down.

Figure 33, Selected Countries' Agricultural Production

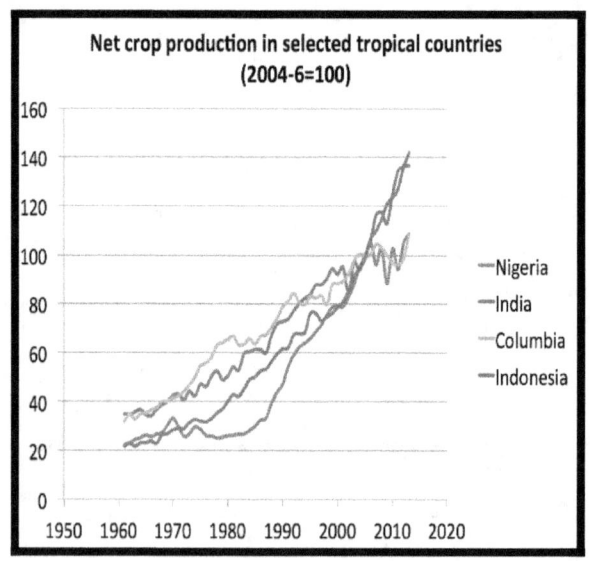

Exploding the Myths of Climate Change
By Tropical Agriculture, Wikipedia, United Nations Food and Agriculture Organization, Statistics Division.

As with the US chart, we see a decided increase in production per acre. Again from the late 1950s to today, we see an increase of 400%-500%. Still, we are growing more food in a time of steadily increasing CO2. **It is not causing droughts and famine, as alleged. Quite the opposite. Food production has reached all-time highs.**

Bad Weather

Again the allegations from the true believers are the increase in CO2 (and to a minor extent other greenhouse gases) will cause climate change will, in turn, cause many more storms and much more severe storms. We have reviewed this topic in the chapter that looked individually at floods, hurricanes, fires, etc. Now, I will help the reader look at the total impact of this 'bad weather increase.'

The following chart shows the ultimate measure of disasters - death. It relates to how the number of deaths resulting from all types of bad weather (storms, hurricanes, floods, high temperatures, etc.) has changed over time. If the true believers are right, we should see a dramatic increase in deaths from floods, heat, hurricanes, etc., as we are 'ravaged by the terrors' of climate change.

Figure 34, Deaths from Bad Weather[126]

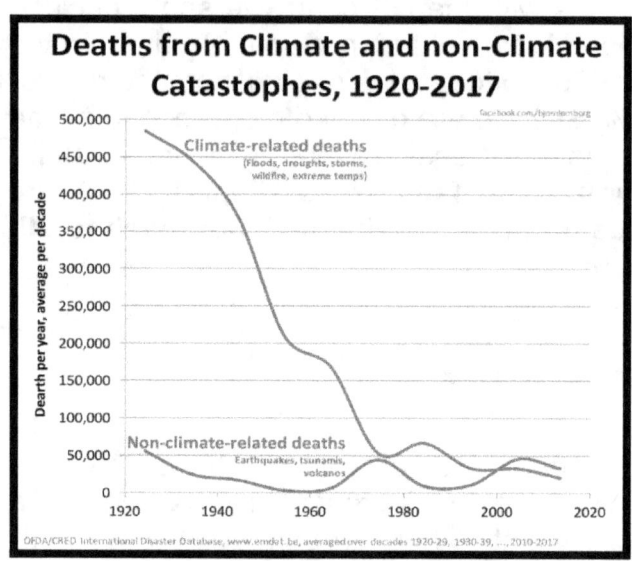

By International Disaster Database

Life Expectancy

Talking about death is the ultimate measure of disasters, how about life expectancy being the ultimate measure of well being? After all, not much is more important. Let's look at life expectancy in the US and the World to see how it is changing.

---

[126] Brandt, Tyler, The Daily Bell, "6 Reasons for Optimism in 2020" - December 30, 2019 – Data from the Brooking Institute.

Exploding the Myths of Climate Change

Figure 35, US Life Expectancy[127]

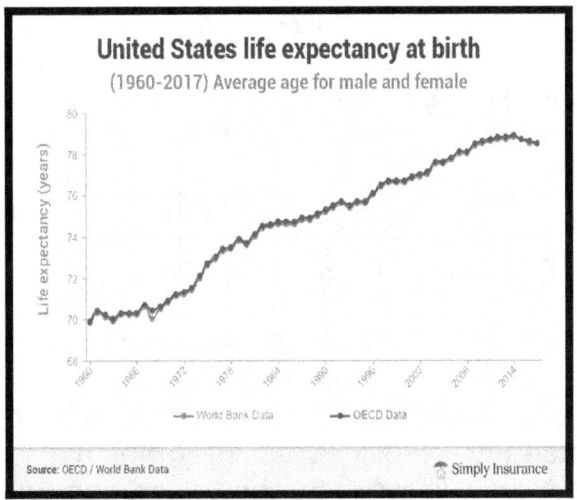

Figure 36, World Life Expectancy[128]

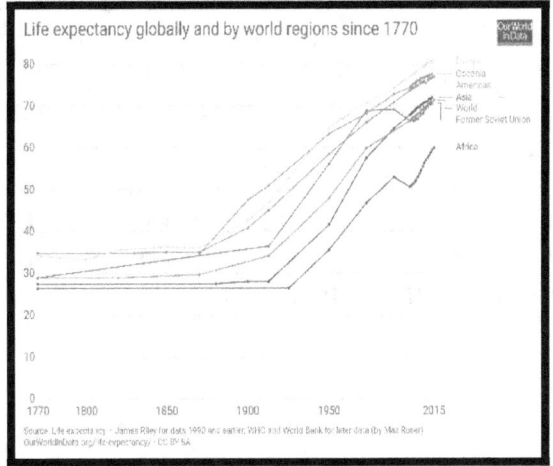

By Our World in Data

---

[127] "US Life Expectancy at Birth," Simply Insurance, https://data.oecd.org/healthstat/life-expectancy-at-birth.htm

[128] Roser, Max, "Life expectancy by World Region, from 1770 to 2018," Our World in Data, October, 2019.

As can be seen from both of these charts, we humans are living longer than ever before, both worldwide and in the US. Again, this is at odds with the 'world is coming to an end' screams we hear from the true believer fanatics.

Summary of Review of Our Real Current Situation

After looking at many important variables, it **should be evident our Earth is a lovely place to live even in this time of steadily rising CO2. People live longer, have more food, are better off economically, and, in general, have a much-improving quality of life.** Let me again say this in no way says everything is perfect. Certainly we are currently (2020-2021) fighting the terrible plague of Covid-19, but we are going to win that fight, even at a high cost. Any amount of poverty is terrible. Any disaster is awful, especially if it happens to you or your loved ones. But on average, **we are notably well off, and our lots in life have been improving, and continue to improve.**

Chapter 11: Exploding the Myth of Climate Change: Summary

The Myths:
- *Global Warming/Climate Change is causing floods, storms, droughts, famines, species extinction, rising seas, melting glaciers to get worse.*
- *The hotter air is heating the seas, causing many disasters.*
- *Global Warming/Climate Change is based on a correlation between rising CO2 and rising temperatures.*
- *97% of all scientists agree on Global Warming.*
- *We don't have anything to lose from following climate change and its ultimate manifestation, the Green New Deal.*
- *The sky is falling; the sky is falling!*
- *Global Warming/Climate Change is causing increasing numbers of, and severity of, all kinds of natural disasters, and it will worsen.*

The Truth:
- *Floods, storms, droughts, famines, species extinction, rising seas, melting glaciers, in fact, are lessening. The worst cases of*

- *most of them occurred decades or centuries before CO2 became an issue.*
- *The physics of thermodynamics does not support the theory hotter air is warming the sea. There is too much water and too little air.*
- *Rising CO2 and rising temperatures don't correlate unless one manipulates a graph to make it appear they do.*
- *97% of scientists don't agree on Global Warming/Climate Change. (97% of scientists probably don't agree on anything!) There is vast disagreement among scientists.*
- *We have everything to lose from the Green New Deal: our way of life, our economy, our country.*
- *The sky isn't falling; Mythmakers over the centuries have sounded false alarms.*
- *Global Warming/Climate Change is not causing a worsening of natural disasters! Natural disasters of all types are at historically low ebb and seem to be getting less frequent and milder when looked at over time.*

We have looked at several aspects of the alleged global warming/climate change issue in this volume. The **main problem with the allegations is there is a lack of cause and effect relationship.** All the true believers claim a range of damaging effects from the increase in CO2, but a close look does not find any real connection. In almost every case, it turns out the worst flood, hurricane, drought, whatever happened long ago, and there is no steadily increasing occurrence of these events or rising of the severity of these events. In judging these impacts, it is important to remain perfectly objective and honest and yet empathetic. All floods, for example, are disastrous. The most recent one is the one most remembered just because it is most recent; that's how humans are wired, and, maybe most importantly, because we now get immediate updates 24X7 from the news media. We tend to forget these bad things have been **happening for many decades, many centuries before the CO2 increased.**

The honesty of reporting is, unfortunately, another issue. We have seen some true believers present pictures of an old, maybe sick, polar bear as evidence (wrongly) a lack of sea ice is destroying the polar bear population. They also like to show pictures of bears looking longingly at the water, as if they can't find ice. (Polar bears are excellent swimmers and can go many miles without a problem, even in the frigid water to the Arctic.) The myth the polar bear population is dwindling (they are doing inordinately well, and their numbers have grown several hundred percent in recent years) has been long since debunked. **However, we still see this intentionally misleading kind of presentation**, usually in an ad **asking for contributions**.

The same is true for the melting sea ice. Again long-debunked, we still see advertisements and articles **(also, usually wanting donations)** explaining how the melting sea ice is causing the oceans to rise and threaten Miami, New York, etc. While the oceans are rising and have been for many decades**, it is not right the melting sea ice** has anything to do with it. As any drinker will know, **ice cubes melting in your drink do not raise the level** in the glass; they just dilute the bourbon in it. The same is true for the sea. Sea ice **melting does not increase its level.**

Many more, somewhat less blatant, misstatements abound in climate change literature. Consider all the many things alleged to be the fault of global warming to see the pattern. I think the media has come to take an easy path by blaming any disaster on global warming/climate change. The 2019 fires in Australia are an excellent example. At first, all the reporting carried a reference to global warming/climate change as the cause or contributing cause. As it became certain the reason was arson, the reports, in this case, at least, stopped blaming global warming.

Let's look at what we have found in the chapters of this book:

- There is **neither pattern of increasing quantity nor severity of weather-** related disasters with increasing CO2.
- Major weather disasters have been befalling mankind back through all recorded history, and the worst generally happened years, decades, or centuries ago. This lack of recentness refutes the argument of climate change (increasing CO2) will cause all kinds of disasters. **The cause and effect relationship is not there.**
- There is a scientific question based on the thermodynamics of water and air regarding the atmosphere's ability to

heat the oceans much. **This question precludes the fundamental logic chain of climate change**, which requires the greenhouse-heated atmosphere to heat the oceans to create many adverse effects.
- There is a **lack of correlation between events and the timing of events** required to support climate change. The Australian wildfires are, again, a good example. As horrific as they are, and as hot as it is there, current temperatures are nowhere close to Australia's hottest, which occurred decades ago.
- The question of **global warming/climate change is crucial, not because we need to take action, but rather because we don't!** The far left in American politics (and in international globalist groups) has grabbed climate change. They are using it as a prod (The Green New Deal) to move toward activities crippling to the United States as we know it. **What we have to lose is our country!**
- Forecasting doomsday events has been a hobby of mankind throughout history. Many examples of such forecasting are **widely believed at the time**, many backed by then-current scientific thought, **but all wrong! Al Gore and AOC are 'superstars' of this group.**
- Despite **all the dire predictions and impacts we should be feeling from climate change, the Earth and its inhabitants are doing remarkably well. Poverty is at its lowest point in history (there is still work to do, of

course); people are better fed; are healthier; live longer; have more money, and have every reason to be happier. <u>We have all this to lose if we follow the excesses recommended by the true believers.</u>

The following drawing presents where we are today on that logic chain.

Figure 37, Real Logic Chain

- There is an increased amount of CO2 in the atmosphere due to the increased use of internal combustion engines and using fossil fuels (coal, oil, and natural gas) for power and heating generation. **Everybody has plenty of power, heat, and cooling.**

- The CO2 is a greenhouse gas, and its increase means more light (heat) may be reflected to Earth rather than being radiated out to space.

- The increased back-reflection may raise air temperatures – **but just a little.**

Exploding the Myths of Climate Change

- The rising air temperatures may cause the seas to heat and raise their temperature – **but just a little at most.**

- The increasing amount of CO2 and maybe air and water temperatures **create an even more pleasant place to live.**

Clifford Holliday

Exploding the Myths of Climate Change
Epilogue

In thinking about what to put in this section as positive steps we should take forward, I have been tempted to say, **"Do nothing." You, true believers, have already done enough damage** with this myth about global warming and climate change.

Despite all the distortions that have been promulgated, our planet:
- Is the healthiest
- Is the richest
- Has the least poverty
- Grows more food
- Has a longer life expectancy
- Has less disease
- Has fewer people dying from disasters

We have achieved all of this while the CO2 has been steadily increasing to record levels. **So don't do anything to mess up this beneficial time**. We seem to be doing just fine with increasing carbon! True, there has been an almost undetectable increase in the average temperature of the atmosphere. Also, there has been an even smaller increase in the temperature of the seas. Some places like the Antarctic Peninsula are heating more than others, but there are always other factors causing these abnormities. Besides, the waters have continued their centuries-long level increase. But there is no 'climate disaster.' All the bad weather and mishaps attributed to climate change have been just in the course of business. In Chapter 3, we saw all of these disasters of today pale by their historical antecedents (when CO2 was much lower.)

So what positive things can we recommend? Here's a list:

1. Keep an active climate research activity going. However, **try not to focus that research on proving an a priori, biased position. Particular emphasis is needed on what is happening at the poles. There is significant disagreement about what is going on in both the Arctic and the Antarctic. Indeed, things are happening in ice melting and ice formation, but the driving forces are not absolute and not understood. To a lesser extent, the same is true as to why the glaciers are retreating. An active research program is also needed in that area.**

2. Maintain an active and vigorous environmental program. (Surprised?) Let's be sure we are actually working on ecological issues - **CO2 is not an environmental issue**; it is a fertilizer and a necessity for life. Smog from particulate matter and smoke is a problem, and we should keep up efforts to require state-of-the-practice remedies such as scrubbers and particulate precipitators. Keeping plastic out of the oceans and numerous other issues are legitimate environmental concerns. **Focus on them. Keep real environmental issues separated from politics.**

3. (This is supposed to be positive, but I just can't help myself with this one.) **Don't confuse global warming with environmental concerns**.

4. All **energy sources are useful and nee**ded – gas, oil, nukes, wind, water, solar, even coal. Let's just work to see they are developed safely and efficiently and used in an environmentally friendly way.

5. Let's work to extend energy to all the people of the globe. Let's boost real, grid-based energy, not just enough for a single candle in the dark. Different locations suggest different types of energy development. For example, if there is abundant hydropower, by all means, let's develop it.

6. Base energy development on the type that best fits the area, not on some preconceived idea about what is 'good' or 'bad.' **All energy that helps man lift himself is good!** We must see it is appropriately developed, provided, and used. **Energy availability is the key to improving everyone's life.**

Clifford Holliday

About the Author – Clif Holliday

In 2001, Clif Holliday decided to write a book about his old teacher and coach, "Tiny" Jones. The result was the Eastern Kentucky classic "Tiny and the Trojans," published in 2002. Since then Clif has spent more and more of his time as a writer. He now has written six books; with five of them published. He writes both fiction and non-fiction. His series of novels all are about the life and times of Alice and Sam and their various adventures. His books are available at:

amazon.com/author/clifholliday

Clif also continues to write major marketing/technical telecommunications reports. These reports are available at:

www.igigroup.com

Clif's recent technical work experience has been focused on the development and publication of major marketing and technical reports in the telecommunications network area. He has also served as a consultant in selected circumstances, founding the consultancy, A & C Consulting Services. In this role he has designed major fiber networks and has been able to assist companies in making design direction and management decisions. These consulting activities have served customers on three continents.

Life in the corporate world began for Clif with a 31 year career with GTE. At GTE he rose to the position of Vice President of Advanced Technology Planning in the Business Development Department. While at GTE, the Governor of Kentucky requested Clif's assistance in the state Energy Department. For well over a year, Clif reported to the Secretary of Energy of Kentucky, as (simultaneously) Director of Demonstration Projects (involving projects worth over $1 B), Director of Energy Studies, and Director of Interface with the Coal Research Laboratory.

EDUCATIONAL BACKGROUND and MEMBERSHIPS

BS EE and MBA (with a Marketing emphasis), University of Kentucky –1963, 1968

Professional Engineer, Kentucky, Texas

Member – IEEE (Life), SPIE (Society of Optical Engineers), National and Texas National and State Societies of Professional Engineers, and many others.

Invited member of Mensa.

Member of the Society of Kentucky Colonels (Awarded for work with the Kentucky Department of Energy.)

Recent Books by Clif Holliday

"Tiny and the Trojans," the story of the life of Tiny Jones, a coach, and teacher.

"Vision Planning – The Key to the Future." A handbook on planning based on desired outcomes in almost any environment.

"Exploding the Myths of Climate Change." This book looks at the many myths sold as truths to support the idea of a man-driven climate change. Issue Two of this effort includes an entire chapter on Texas 'Big Freeze' in 2021 - its causes and needed cures.

"At Your Service – The Faces of Service." This non-fiction book is a handbook for how to achieve 'Great Service' in any organization. It is co-authored with Alice Holliday. The book includes many examples of excellent service, good service, and lousy service.

"The Treasure of the Mount – A Hunt Across Time." This is a bi-era historical fiction novel. It is centered in the 1860s on a robbery incident late in the Civil War. The current period is concerned with the rekindling of a high-school romance between Alice and Sam.

"Electric Cars and You." This non-fiction provides all the detail you what to know about the much talked about electric cars. This

Exploding the Myths of Climate Change book will give you all the technical information and the pros and cons of electric vehicles. It also includes extensive treatment of the economics of EVs and various listings of those available in each type.

(All above books are available on Amazon at: amazon.com/author/clifholliday)

"Late Love and the Lost Gold." This is the second in the series of Alice and Sam books. It is a historical fiction concerning the last days of the Confederacy and a treasure that is lost leaving Richmond. Presently, Sam and Alice struggle to keep their relationship alive and stumble into a dangerous treasure hunt.

"Rat-a-tat-tat!" This further adventures of Alice and Sam book is in the character of a historical novel, but it is set a few years in the future. It is centered on the two lovers' involvement in a presidential campaign.

"Flying Saucers, Black Cats, and Promethium." This is the latest in the Alice and Sam series. In this novel, set a few years in the future, but taking headlines from today, Sam is cursed by a Black Cat, and the president Charges the two lovers with discovering the secrets of the flying saucers.

These last three are looking for a publisher.

www.ingramcontent.com/pod-product-compliance
Lightning Source LLC
Chambersburg PA
CBHW071400210526
45465CB00001B/185